Understanding Radiation Biology

From DNA Damage to Cancer and Radiation Risk

T0332046

Understanding
Radiation Biology

From DNA Damage to Cancer
and Radiation Risk

K. H. Chadwick

CRC Press
Taylor & Francis Group
Boca Raton London New York

CRC Press is an imprint of the
Taylor & Francis Group, an **informa** business

CRC Press
Taylor & Francis Group
6000 Broken Sound Parkway NW, Suite 300
Boca Raton, FL 33487-2742

International Standard Book Number-13: 978-0-367-25376-9 (Paperback)
International Standard Book Number-13: 978-0-367-25515-2 (Hardback)

Library of Congress Cataloging-in-Publication Data

Names: Chadwick, K. H. (Kenneth Helme), 1937- author.
Title: Understanding radiation biology : from DNA damage to cancer and radiation
risk / by Kenneth Chadwick.
Description: Boca Raton: CRC Press, [2020] | Includes bibliographical references and index.
Identifiers: LCCN 2019031202 (print) | LCCN 2019031203 (ebook) | ISBN 9780367253769
(paperback) | ISBN 9780367255152 (hardback) | ISBN 9780429288197 (ebook)
Subjects: LCSH: Radiobiology.
Classification: LCC QH543.5 .C44 2020 (print) | LCC QH543.5 (ebook) | DDC 571.4/5--dc23
LC record available at https://lccn.loc.gov/2019031202
LC ebook record available at https://lccn.loc.gov/2019031203

Visit the Taylor & Francis Web site at
http://www.taylorandfrancis.com

and the CRC Press Web site at
http://www.crcpress.com

To my wife, Hilary, my daughters, Carolyn and Victoria,
and my Bichon Frisé, Bonnie

Contents

PART II   Ultraviolet Light Effects

PART III   Genotoxicology

Prologue

The aim of this book is to present a coherent, qualitative and quantitative theory of radiation biology which has been developed and expanded over the years from the 1970s to present day. Many readers of the wide variety of separate scientific articles, published as part of this development, will probably not realise that a comprehensive theoretical model lies behind them, which provides in-depth understanding of radiation biology. Consequently, I felt that it would be a useful exercise to bring together all the current concepts and ideas arising from this model in one place, this new book.

In 1971, my colleague, Dr H. P. Leenhouts, and I were examining cell survival curves to see if we could find any mathematical consistency in them and noticed that, when we plotted the logarithm of survival against the square of the radiation dose, the data invariably lay on a straight line which crossed the zero dose axis, just under the origin of 100% cell survival (see Figure P.1).

We soon realised that a linear–quadratic dose–effect equation of the type:

$$S = \exp\left(-p\left(aD + bD^2\right)\right)$$

where (S) is cell survival, (D) is radiation dose and (a) and (b) are coefficients that can be determined by fitting the equation to the data, provided, to our rather inexperienced eyes, a very good description of a large number of cell survival data published by others in the scientific literature (see Figure P.2 for an example).

We later discovered that, in 1966, Sinclair (1966) had found, by using a computer to fit a selection of different mathematical functions to his cell survival data, that the linear–quadratic equation gave the optimum fit. To the best of our knowledge, Sinclair did not follow up on this work.

Our further analysis of published data revealed that the coefficient (a) depended strongly on radiation quality, increasing for more densely ionising radiation, and that the coefficient (b) depended on radiation dose rate, decreasing as the exposure was protracted. There was a good consistency in the analyses we made and the equation offered interesting perspectives.

We then ran into a quandary. What was the basis for the linear–quadratic equation? One of us (KHC) had had some training in radiation biology and was aware that the Classical Theory of chromosome aberration formation proposed a linear–quadratic equation to describe the dose–effect curves of aberration yield and we therefore spent some time debating the question, 'Is it simply chromosome aberrations that give cell inactivation?' We found ourselves going round in a circle, not getting any further. Fortunately, the other one of us (HPL) came from the field of nuclear physics and was not instilled with radiation biology concepts. He was able to think outside the circle and, after considerable discussion, we concluded that, from a purely hypothetical point of view, the double stranded nature of the DNA molecule in the nucleus of the cell would provide a logical target and that radiation-induced double strand breaks in the DNA could, potentially, have a linear–quadratic dose–effect relationship.

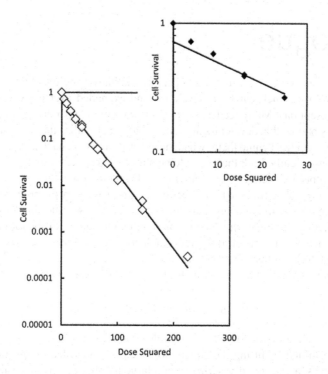

FIGURE P.1   Cell survival data plotted against the square of the radiation dose. The inset shows the detail close to 100% cell survival and the straight line crossing zero dose at 72% cell survival.

It was only when we turned away from the Classical Theory of chromosome aberration formation to assume that radiation-induced DNA double strand breaks might be induced in proportion with the dose and with the dose squared and, thus, be responsible for cell inactivation, that we were able to move forward. A whole panoply of radiation biological phenomena which had puzzled us suddenly started to find logical explanations and one revelation led to another. Each conclusion conjured up a new proposition and different cellular endpoints could be traced to the same type of lesion, correlations were predicted and found. We started to see through the mist and gradually developed an insight into radiation biological action.

Our assumption was not, of course, without its problems. In proposing that DNA double strand breaks were the critical radiation-induced lesion, we tied ourselves to what was known about the DNA and double strand breaks in the cell in 1971, which was not a great deal, and we also tied ourselves to what would be discovered about DNA and double strand breaks from 1971 onwards.

In 1971, we were complete novices in the interpretation of radiobiological data but, as we developed the double strand break model, we rapidly found that a clear and novel vista was opening and our insights, while in contradiction with conventional dogma, offered interesting perspectives. It was as if a jigsaw puzzle was being put together bit by bit to create a complete picture and, within ten years, we had written a book outlining the earlier part of our work (Chadwick and Leenhouts, 1981).

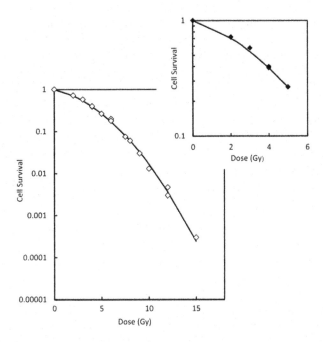

FIGURE P.2   The same cell survival data as shown in Figure P.1 plotted against the radiation dose. The inset shows the detail close to 100% cell survival and the line is given by the linear–quadratic equation: $S = \exp(-(0.12D + 0.029D^2))$.

In spite of the 1981 book, the various bits which make up the complete jigsaw puzzle are scattered in scientific publications that are not always easily accessible and, in any case, many post-date the 1981 book. In addition, some proposals presented in that book, such as one on the formation of chromosome aberrations and another on radiation-induced cancer, have been superseded and we have refined and redefined them.

The main thrust of our work has been concerned with the development of a straightforward but flexible model to provide a comprehensive and coherent, quantitative description of radiation effects. The severity of the radiation biological effect depends on many factors which embrace the physics of energy deposition, the chemistry of sensitising and protective cellular environments, free radical chemistry, the biochemistry and genetics of DNA repair, and the structure and behaviour of the DNA target molecule in the cell cycle. In view of this, it is unlikely that any theoretical model will provide a complete explanation of all the different aspects of radiation biology. However, the model we have derived does take account of the physical, chemical and biological parameters which influence the subsequent radiation effect. It develops a mathematical description of dose–effect relationships which can be used to analyse and interpret experimental data, provides links between different cellular endpoints and has predictive qualities relevant for radiological protection. The quantitative cellular model lends itself to application in an established model of cancer induction and has been used to interpret radiation-induced cancer in animals and man with illuminating predictions for radiation risks. Finally, the model has been expanded to consider the effects of ultraviolet radiation and genotoxic chemicals. In all these different areas, the model gives deep insight into the basic action of radiation at the cellular level as well as offering important predictions on genotoxic cancer risk.

This book reviews most of our relevant publications and sets them in the broader scope of modelling the action of all genotoxic agents. It presents some unpublished data and analyses, and it outlines, in more detail than has been possible in individual publications, a myriad of thoughts and considerations that the development of the model stimulated.

Since the 1980s, the linear–quadratic equation has been widely used, both in radiobiology and in the radiotherapy field, to analyse radiation dose effect data as it invariably gives a good fit (Brenner 2008; Astrahan 2008; McKenna and Ahmad 2011; Franken et al. 2013; Santiago et al. 2016; Van Leeuwen et al. 2018). Although there are several different approaches to the derivation of the equation (Lea and Catcheside 1942; Kellerer and Rossi 1972, 1978; Chadwick and Leenhouts 1973a, 1981; Herr et al. 2015; Bodgi and Foray 2016; Bodgi et al. 2016), the analyses have often been made without much reference to the differing interpretations offered by the different derivations. This is unfortunate, as it is the combination of the data analysis with a mechanistic interpretation which provides insight. I hope that this book will lead the reader to consider the analytical application of the linear–quadratic equation in terms of the molecular lesion, the DNA double strand break, proposed here as the crucial lesion which leads to the wide variety of radiobiological effects and discover an insight into, and an understanding of, the biological action of radiation.

Acknowledgements

I wish to record the immense satisfaction that I have experienced in the friendly but very professional interaction with all the staff at CRC Press/Taylor & Francis, especially Kirsten Barr and Rebecca Davies, and all at Deanta Global, especially Conor Fagan. Without their help, expertise and advice during the preparation of this book, it would not have been possible to bring it to completion. It has been a pleasure to work with them. Much of the research presented here was supported by the Euratom Radiation Research Programme of the European Commission and the Dutch Ministries of Agriculture and of Public Health and the Environment.

K. H. Chadwick
Kendal, United Kingdom

Part I

Ionising Radiation Effects

1 The Molecular Model and DNA Double Strand Breaks

A detailed, parameterised linear–quadratic dose–effect equation for the induction of DNA double strand breaks by ionising radiation is derived based on the known structure and properties of the DNA molecule in eukaryotic cells. Experimental measurements of the dose–effect relationships for the induction of DNA double strand breaks are presented, in support of the linear–quadratic dose–effect equation. The inferences, implications and insights for radiation action which can be drawn from a consideration of the detailed linear–quadratic equations are elaborated.

1.1 THE MOLECULAR MODEL HYPOTHESIS – THE BASIC CONCEPTS

The molecular model used to provide a qualitative and quantitative description of ionising radiation effects in cells is based on just two postulates:

1. The DNA double strand break is the crucial cellular lesion which may lead to cell inactivation, chromosomal aberrations and mutations.
2. The dose–effect relationship for the induction of DNA double strand breaks is linear–quadratic.

    Everything else derives directly from these two postulates as straightforward consequences which depend on the structure and properties of the DNA molecule, on the chemical surroundings of the DNA, and on the track structure and properties of the different radiations.

A third postulate is:

3. The radiation effects induced in cells lead to the various radiation-induced health effects.

The first postulate, that the double strand break is a critical lesion for a cell, is not contentious and is generally accepted, but there are two major objections to the model. One, which can be called the 'micro-dosimetry problem', concerns the probability that two independently induced DNA single strand breaks will be close enough to create a double strand break at radiation doses relevant for the biological effects. The other objection, which can be called the 'cytological problem', concerns the production of chromosome exchange aberrations from a single DNA double strand break.

These two objections will be addressed at the appropriate stages as the development of the model is expanded through the book.

In this first chapter, a detailed, parameterised linear–quadratic dose–effect equation for the induction of DNA double strand breaks by ionising radiation is derived using the known structure and properties of the DNA molecule in eukaryotic cells. Experimental measurements of the dose–effect relationships for the induction of DNA double strand breaks are presented confirming the linear–quadratic dose–effect equation. The equation is then used in Chapter 2 to develop dose–effect relationships linking the number of double strand breaks to three cellular effects: cell death, the yield of chromosomal aberrations and mutation frequency.

1.2   THE INDUCTION OF DNA DOUBLE STRAND BREAKS

There are good reasons for choosing deoxyribonucleic acid (DNA) as the important target for radiation effects. DNA is common to all living cells and provides the universal genetic code. It has a high molecular weight and forms the backbone of the chromosomes which are contained in the nucleus of the cell. The DNA in a cell controls the internal working and defines the specific activity of the cell in an organism. Any disruption of the mechanical or genetic integrity of the DNA molecule will clearly have serious consequences for the continued normal function of a cell.

The DNA molecule has a well-defined three-dimensional structure originally determined by Watson and Crick in 1953 (Watson and Crick 1953). Two long polymer chains of alternating sugar and phosphate units are wound around each other in the form of a double helix. The two sugar-phosphate polymer chains are linked at each sugar unit by one of two purine–pyrimidine pairs of nucleotide bases, adenine–thymine pairs (A–T) and guanine–cytosine pairs (G–C), and, because the dimension of the A–T pair is the same as the dimension of the G–C pair, the two sugar-phosphate chains are held parallel to each other, separated by 1.2 nm, so that the structure resembles a long, twisted, lightly coiled, rope ladder on a molecular scale. The two sugar-phosphate strands are wound round each other to make one full turn every 3.4 nm in a right-handed spiral, which in turn is wound around a central axis so that a major groove and a minor groove are formed. Each complete unit of base plus sugar plus phosphate is called a nucleotide so that each strand of the DNA is a polynucleotide chain. The nucleotide purine–pyrimidine pairing occurs every 0.34 nm along the sugar-phosphate chain so that there are ten links holding the chain together for every full turn of the spiral. The sequence of the purine (A, G) and pyrimidine (C, T) bases along the chain forms the basis of the genetic code. The result, illustrated in Figure 1.1, is a very long, thin molecule reaching up to 50 mm in length with a diameter of 2 nm.

In addition to providing the basis of the genetic code, the complementary base pairing makes it possible for the DNA molecule to replicate itself correctly during the DNA synthesis (S) phase of the cell cycle. In DNA synthesis, the two 'old' strands of DNA loosen and replication starts at many replication origins, proceeding in both directions along the DNA (Benbow et al. 1985; Linskens and Huberman 1990; Douglas et al. 2018). The 'old' strands are copied to make two 'new' strands with complementary base pairing so that the two new double helices are exact copies

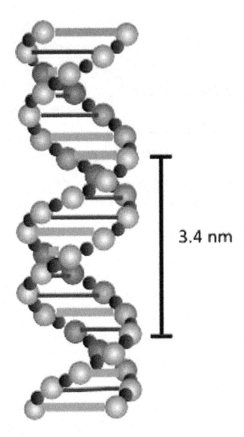

3.4 nm

FIGURE 1.1   A schematic representation of the DNA double helix molecule.

of the original double helix and each of the two helices has one 'old' strand and one 'new' strand (see Figure 1.2). At mitosis, the two new DNA double helices separate into two daughter cells, each of which carries the same genetic information from the original cell.

1.3   THE LINEAR–QUADRATIC FUNCTION AND DNA DOUBLE STRAND BREAKS

It is not difficult to understand how a linear–quadratic dose–effect relationship for the induction of double strand breaks, which obviously disrupt the integrity of the DNA double helix molecule, can be derived from the interaction of radiation tracks with the molecular structure presented in Figure 1.1. A double strand break can be induced as a consequence of one ionising radiation track breaking both strands of the DNA double helix, giving a yield of breaks in proportion with radiation dose (αD). A DNA double strand break can, at least hypothetically, also result as a consequence of the close spatial proximity of two independently induced single strand breaks, giving

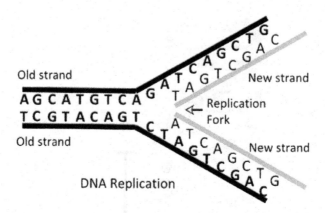

FIGURE 1.2 A schematic diagram of the process of the replication of DNA.

a yield of double strand breaks in proportion with the square of the radiation dose (βD^2), as is illustrated in Figure 1.3.

In accordance with these two modes of radiation action, the average number (N) of DNA double strand breaks per cell induced by a dose (D) of radiation is, in general, given by the equation:

$$N = \alpha D + \beta D^2. \tag{1.1}$$

Figure 1.4 presents the number (N) of double strand breaks as a function of dose (D) according to the linear–quadratic Equation 1.1, broken down into its two components to show that the (α) coefficient represents the initial linear slope of the curve from zero dose, and the (β) coefficient accounts for the upwards bending of the curve.

Knowledge of the structure of the DNA double helix and its properties in the cell, as well as knowledge of the structure of ionising radiation tracks permits us, but also obliges us, to define the (α) and (β) coefficients in considerable detail by including parameters to cover all the different processes which can be logically considered to be involved in the induction of strand breaks.

1.4   THE ALPHA MODE OF DOUBLE STRAND BREAK INDUCTION

A careful consideration of the alpha mode of double strand break induction shown in Figure 1.3, intuitively leads to a detailed derivation of the (α) coefficient by taking a series of parameters into account which influence the probability that a DNA double strand break is induced in the passage of a single ionising particle track.

If

- n is the number of nucleotide base pairs per cell representing the amount of DNA in the nucleus of the cell which is the target molecule. (*In a human cell, the number of nucleotide base pairs (n) is approximately* 3×10^9.)
- μ is the probability per unit dose that an ionising particle passes close to a nucleotide base. (*This parameter is dependent on the flux of particle tracks,*

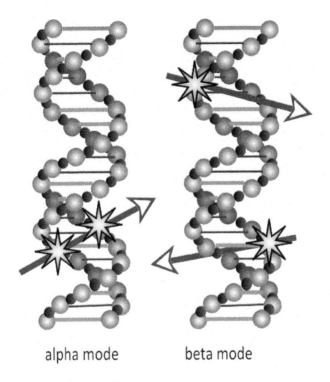

alpha mode   beta mode

FIGURE 1.3 Schematic representation of the hypothetically possible formation of DNA double strand breaks in two different modes of radiation action.

usually secondary electrons, generated by and dependent on the type of incident radiation.)

- k is the probability per nucleotide base that, when an ionising particle passes close to a nucleotide base, an energy deposition occurs which eventually leads to a strand break most probably as a result of radical attack. *(This parameter is dependent on the chemical environment around the DNA and, more specifically, on the radical scavenging in the nucleus of the cell.)*

- Ω is the probability that when the ionising particle passes close to the 'first' strand it also passes close to the 'second' strand. *(This parameter clearly depends on the spatial energy deposition pattern along the particle track and, thus, on the type of incident radiation, as well as the particular three-dimensional structure of the DNA molecule, especially the 1.2 nm distance between the two sugar-phosphate strands.)*

So that

- Ωk represents the probability per 'first' strand break that the 'second' strand is broken in the passage of the same ionising particle by a second energy deposition and radical attack. *(The implication is that the energy*

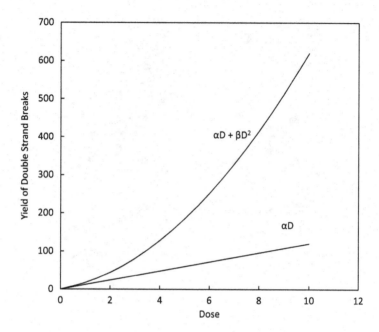

FIGURE 1.4 The yield of DNA double strand breaks (arbitrary units) as a function of dose according to Equation 1.1.

depositions have to be close to each strand and that two radicals, one for each strand break are involved. Prise et al. [1993, 1999] have demonstrated that the double strand break originates from two radicals.)

Then, the (α_0) coefficient can be derived as

$$\alpha_0 = 2n\mu k\Omega k \tag{1.2}$$

where the factor 2 merely indicates that either strand can be 'first', and the number (N_1) of DNA double strand breaks induced in the alpha mode of radiation action by a dose (D) is given by:

$$N_1 = \alpha_0 D = 2n\mu k\Omega kD. \tag{1.3}$$

1.5   THE BETA MODE OF DOUBLE STRAND BREAK INDUCTION

A similar consideration of the beta mode of double strand break induction, shown on the right of Figure 1.3, leads to a detailed derivation of the (β) term by taking a series of parameters into account. These parameters influence the probability that a DNA double strand break is induced as a consequence of the close proximity of two independently induced single strand breaks. The derivation follows from the previous reasoning, making use of the same parameters.

If

- $(1 - \Omega)$ is the probability that when the ionising particle passes close to the 'first' strand, it does not pass close to the 'second' strand

and

- $\Omega(1 - k)$ is the probability that when the ionising particle passes close to the 'first' strand, it also passes close to the 'second' strand but does not give an energy deposition leading to a break in the 'second' strand

then

- $(1-\Omega)+\Omega(1-k)=1-\Omega k$, is the total probability per 'first' strand break that the 'second' strand is not broken in the passage of the same ionising particle.

As

- $n\mu k$ is the probability per cell per unit dose that a break occurs in the 'first' strand,

then

- $2n\mu k(1-\Omega k)$ represents the probability per cell per unit dose that a single strand break occurs in the 'first' strand but that it is not accompanied by a break in the 'second' complementary strand in the passage of the same particle.

Again, the factor 2 merely indicates that each strand can be 'first'.

These 'first' single strand breaks can be converted into a double strand break if a 'second' single strand break is induced in the complementary strand of the DNA as a result of an energy deposition event caused by a different ionising particle in the neighbourhood of the 'first' single strand break before it has been repaired.

So, if

- n_1 is the number of nucleotide bases opposite the 'first' single strand break within which a 'second' break will convert the 'first' single strand break to a double strand break,

then

- $n_1\mu_1 k_1$ represents the probability per unit dose per 'first' single strand break that a 'second' break is induced in the passage of a separate ionising particle, and the number (N_2) of double strand breaks induced in this mode of radiation action is given by:

$$N_2 = \beta_\infty D^2 = 2n\mu k(1-\Omega k)n_1 \mu_1 k_1 (D^2/2) \qquad (1.4)$$

where the coefficient (β_∞) is the maximum value of the quadratic coefficient after an acute exposure when there is no repair of the 'first' single strand breaks.

The term $(D^2/2)$ arises because, although the number of 'first' single strand breaks is proportional to dose (D), the 'second' single strand break can only be made when the 'first' is already present so that, on the average, the probability for the 'second' break is proportional to one-half of the dose (D/2).

A distinction has been made between the parameters (μk) involved in the induction of the 'first' breaks and the parameters ($\mu_1 k_1$) involved in the induction of the 'second' breaks because it is not certain that the same mechanism (radical species) is involved in this second mode of radiation action. This point is important as it impacts on the contentious nature of the quadratic term of the linear–quadratic equation that has been derived for the induction of DNA double strand breaks.

Consequently, the total number (N_T) of DNA double strand breaks induced by a dose (D) of radiation is given by the sum of Equations 1.3 and 1.4, though it has to be noted that no account has yet been taken of repair processes:

$$N_T = \alpha_0 D + \beta_\infty D^2 = 2n\mu k\Omega kD + n\mu k(1-\Omega k)n_1\mu_1 k_1 D^2. \qquad (1.5)$$

1.6   THE INFLUENCE OF REPAIR

The repair of both DNA single strand breaks and DNA double strand breaks can influence the number of double strand breaks which are induced and remain biologically effective. Both repair processes depend on the metabolic activity of the cell, the cell stage and the time available for repair. The repair of single strand breaks is efficient and error-free because the nucleotide bases on the undamaged DNA strand can be used by the repair enzymes to reconstruct the broken strand, according to the complementary base-pairing rules. The repair of double strand breaks, a process that was barely hinted at when the model was first derived in the early 1970s (Chadwick and Leenhouts 1973a), is now thought to be achieved via two main pathways, non-homologous end joining (NHEJ) and homologous recombination (HR) and, although both processes may restore the mechanical integrity of the DNA, the genetic integrity of the DNA may not be completely restored. This means that the repair of DNA double strand breaks cannot be guaranteed to be error-free.

1.6.1   SINGLE STRAND BREAK REPAIR

Further consideration of the two modes of double strand break induction reveals that the two energy deposition events from the same particle track in the alpha mode must occur simultaneously so that there will not be any exposure time–related effects on this mode of radiation action. However, the two energy deposition events from two different particle tracks in the beta mode are not necessarily simultaneous. Thus, when the time between the 'first' single strand break and the 'second' break is long enough, the 'first' break can be repaired correctly before the 'second' break is induced. This single strand break repair affects the beta mode of double strand break induction and in a sub-acute exposure can lead to a reduction of the (β) coefficient although the (α) coefficient is unaffected.

If

- f_1 is defined as the proportion of 'first' single strand breaks which are not repaired before the 'second' break converts it to a double strand break,

then the quadratic term for the number of DNA double strand breaks becomes:

$$f_1 N_2 = f_1 \beta_\infty D^2 = f_1 n\mu k(1 - \Omega k) n_1 \mu_1 k_1 D^2 \tag{1.6}$$

and the number (N_0) of initially induced DNA double strand breaks is given by:

$$N_0 = \alpha_0 D + \beta_0 D^2 = 2n\mu k\Omega kD + f_1 n\mu k(1 - \Omega k) n_1 \mu_1 k_1 D^2. \tag{1.7}$$

The parameter (f_1) can have values between 0 (*full repair*) and 1 (*no repair*).

It is important to note that, in the model, the 'second' single strand break converts the 'first' single strand break to a double strand break at the moment that the 'second' break occurs.

1.6.2 DOUBLE STRAND BREAK REPAIR

There is no reason to assume *a priori* that the processes involved in the repair of DNA double strand breaks can distinguish between those induced in the alpha mode or in the beta mode of radiation action and therefore this repair is included as follows:

If

- f_0 is the proportion of double strand breaks which are not repaired and remain biologically effective,

then the number of induced and effective double strand breaks (N) is given by:

$$N = f_0 N_0 = \left(f_0 2n\mu k\Omega kD\right) + f_0 f_1 n\mu k\left(1 - \Omega k\right) n_1 \mu_1 k_1 D^2 \tag{1.8}$$

and

$$N = \alpha D + \beta D^2. \tag{1.1}$$

The parameter (f_0) may have values between zero and one and depends on the metabolic activity of the cell and on the time between the induction of the double strand breaks and the moment when they become biologically effective, which is suspected to be at the first mitosis following exposure. It seems highly unlikely that all the double strand breaks will be perfectly repaired and that (f_0) will be zero. There may be situations or cell types where none of the double strand breaks are repaired when (f_0) would be one, so in general, (f_0) is expected to have a value less than unity.

The repair parameters (f_1) and (f_0) are assumed to have constant values within an experiment and be independent of the radiation doses applied. There is some evidence for this as Dugle and Gillespie (1975) have reported that, in Chinese hamster cells, there was no difference in the repair rate constant for DNA single strand breaks in the dose range from 40 to 400 Gy (*Gray is the unit of radiation dose*). In the case of DNA double strand breaks, Corry and Cole (1973) found no change in repair in Chinese hamster ovary cells up to 500 Gy, and Resnick and Martin (1976) reported that a similar proportion of breaks were repaired in yeast cells in the dose range from 250 to 1000 Gy.

Equation 1.8 provides a detailed elaboration of the (α) and (β) coefficients of Equation 1.1 using information on the DNA helix molecular target, the double strand break lesion and the track structure of radiation.

1.7 THE EXPERIMENTAL EVIDENCE FOR THE LINEAR–QUADRATIC DOSE RESPONSE OF RADIATION-INDUCED DNA DOUBLE STRAND BREAKS

Although the accurate measurement of DNA single strand breaks was well established when the model was developed in 1973, the measurement of DNA double strand breaks at doses relevant for radiation biology was not possible until the 1980s.

The first indication of a quadratic component in the dose–effect for the induction of DNA double strand breaks was found by Dugle, Gillespie and Chapman (Dugle et al. 1976) using the measurement of both DNA single strand breaks by alkaline gradient sedimentation and DNA double strand breaks by neutral gradient sedimentation in quite heavily irradiated cells after incubation for repair. Dugle, Gillespie and Chapman found that after incubation for repair there were two residual single strand breaks for every residual double strand break at each dose. They also showed that both the residual single and double strand breaks were directly related to the square of the radiation dose. Figure 1.5 presents their strand break data redrawn as a function of dose in order to emphasize the curvature of the relationship between DNA double strand breaks and radiation dose. The radiation doses are indeed well beyond the range which is of interest for cellular radiation biology but the curvature of the dose response found for the induction of DNA double strand breaks is significant and Dugle, Gillespie and Chapman concluded that the coefficient for double strand breaks, induced by two independent particle events, is large enough to support the model developed in Sections 1.1 to 1.6.

In 1982, Dikomey (1982) used a similar experimental strategy to that of Dugle et al. (1976) measuring residual DNA single strand breaks after a one hour repair period post exposure, assuming that the single strand break measurement was a reliable measurement of DNA double strand breaks. Dikomey was able to measure the DNA double strand breaks in this way at doses which were relevant for cellular radiation biology and used three different X-ray and hyperthermia treatments of Chinese hamster ovary (CHO) cells. In all three cases, a linear–quadratic relationship between the number of DNA double strand breaks and X-ray dose was found as shown in Figure 1.6.

In the mid-1980s, it became possible to measure DNA double strand breaks directly at radiation doses relevant to cellular radiation biology using neutral gradient filter elution techniques. In 1985, Radford in Australia, was the first to publish a

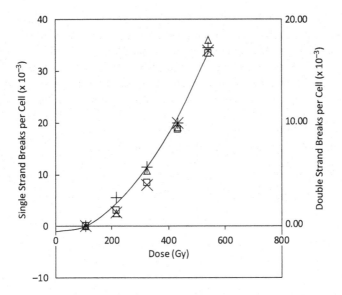

FIGURE 1.5 The yield of single and double strand breaks measured after incubation for repair as a function of dose. The open symbols are for single strand breaks after incubation for 210 and 310 minutes and relate to the left scale. The crosses are for double strand breaks after incubation for 210 and 310 minutes and relate to the right scale (data from Dugle et al. 1976).

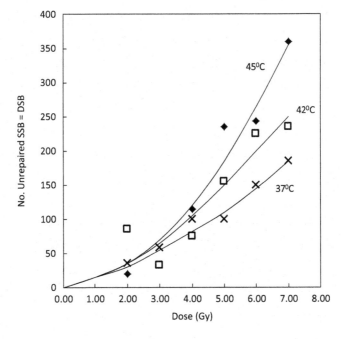

FIGURE 1.6 The linear–quadratic relationship between the number of DNA double strand breaks and X-ray dose after three hyperthermia treatments in CHO cells (data from Dikomey 1982).

series of direct measurements of DNA double strand breaks as a function of X-ray dose in mouse L cells using neutral filter elution at doses in the biologically relevant range (Radford 1985). Radford exposed the mouse L cells to radiation under several different exposure conditions such as hyperthermia, radiation sensitisers and radiation protectors. Although several different response curves were measured, in all the different treatments the dose–effect relationship for the induction of DNA double strand breaks was shown to be linear–quadratic (Radford 1985). These data are presented for clarity in two separate (see Figures 1.7 and 1.8).

Radford also demonstrated that the linear–quadratic relationship between DNA double strand breaks and radiation dose was not just restricted to mouse L cells by making comparable measurements in human, Chinese hamster cells and another mouse cell line (Radford 1986). Figure 1.9 presents, as an example, the data for human cells irradiated in air and Chinese hamster V79 cells irradiated in air and with the protector cysteamine.

In 1987, Prise, Davies and Michael (Prise et al. 1987) in the United Kingdom, used the neutral elution method to measure the induction of DNA double strand breaks as a function of X-ray dose in Chinese hamster cells in air and under hypoxia. Their results are presented in Figure 1.10.

Prise et al. (1987) also made measurements of DNA double strand breaks after irradiation of the cells with neutrons and alpha particles and later (Prise 1994), with protons of differing energies producing dose–effect relationships which varied from linear–quadratic to purely linear in form. These results are discussed later.

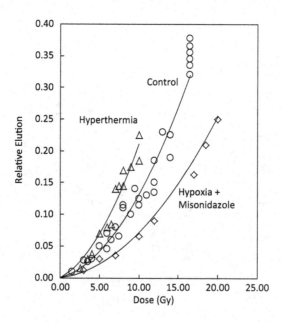

FIGURE 1.7 The linear–quadratic relationship between the number of DNA double strand breaks (relative elution) and radiation dose after three different treatments in mouse L cells (data from Radford 1985).

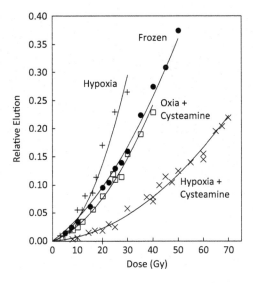

FIGURE 1.8 The linear–quadratic relationship between the number of DNA double strand breaks (relative elution) and radiation dose after four more different treatments in mouse L cells (data from Radford 1985).

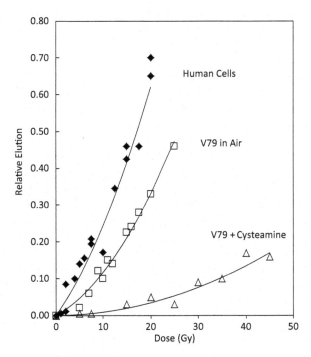

FIGURE 1.9 The linear–quadratic relationship between the number of DNA double strand breaks and radiation dose in human and V79 hamster cells in air and V79 hamster cells with cysteamine (data from Radford 1986).

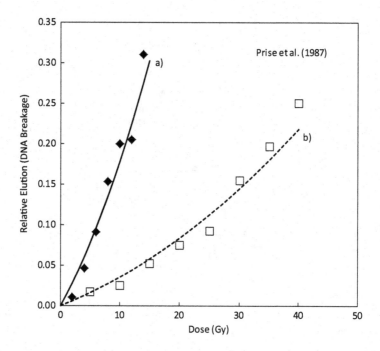

FIGURE 1.10 The linear–quadratic relationship between the number of DNA double strand breaks (relative elution) and X-ray dose in Chinese hamster cells irradiated in air (a) and under hypoxia (b) (data from Prise et al. 1987).

A third group, Murray, Prager and Milas (Murray et al. 1989) in the United States, used the neutral filter elution method to investigate the protective effect of amino-thiols on the induction of DNA double strand breaks after gamma irradiation. Their results are presented in Figure 1.11.

Other groups have also studied the induction of DNA double strand breaks as a function of radiation dose, but usually at higher doses than those illustrated in the Figures 1.7 to 1.11. In general, similar curved dose responses were found for the DNA double strand breaks in these higher dose studies (Sweigert et al. 1988; Peak et al. 1991b) as can be seen in Figures 1.12 and 1.13.

Clearly, there are several measurements of radiation-induced DNA double strand breaks, using either a direct method of neutral filter elution or via an indirect measurement of unrepaired DNA single strand breaks, all of which report linear–quadratic shaped dose–effect responses in agreement with the second postulate of the molecular model outlined in Section 1.1 and with the simple linear–quadratic Equation 1.1 and the more elaborated Equation 1.8.

1.8   INFERENCES, IMPLICATIONS AND INSIGHTS

When the equation for the induction of DNA double strand breaks as a function of radiation dose was initially derived and applied to the analysis of cell survival, in 1972 (Chadwick and Leenhouts 1973) an attempt was made to include separate parameters

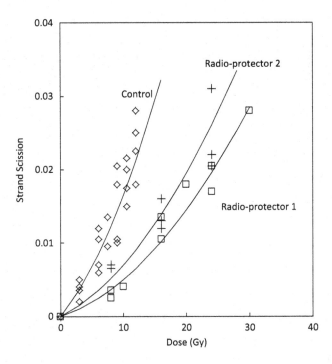

FIGURE 1.11 The linear–quadratic relationship between the number of DNA double strand breaks and radiation dose in CHO cells after exposure in air and with two radio-protectors (data from Murray et al. 1989).

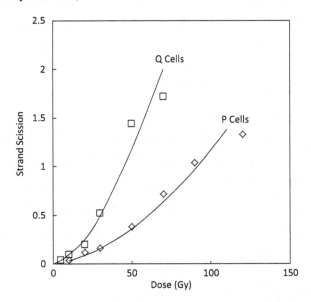

FIGURE 1.12 The curved relationship between the number of DNA double strand breaks and radiation dose in proliferating (P) and quiescent (Q) mouse mammary cells (data from Sweigert et al. 1988).

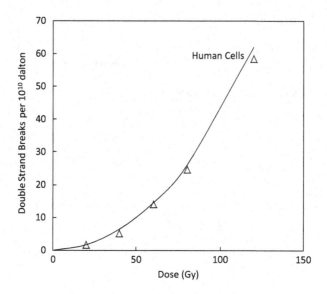

FIGURE 1.13   The curved relationship between the number of DNA double strand breaks and radiation dose in human epithelioid cells (data from Peak et al. 1991b).

to cover all the different processes which could be imagined to be involved. This led to the detailed formulation of the two modes of radiation action outlined in Equation 1.8. Using this detailed formulation of the two modes of radiation action, it is possible to draw some inferences about the effect of radiation at the molecular lesion level which, with the extension of the model to include cellular radiation effects such as cell inactivation, chromosomal aberrations and mutations, have implications for radiation biology and radiological risks.

1.8.1   The Alpha Mode of Radiation Action

$$\alpha = f_0 N_1 = 2 f_0 n \mu k \Omega k D \tag{1.9}$$

The concept of an ionising particle track interacting with a three-dimensional target, the DNA double helix, has been incorporated into the (α) term, using the parameters (μ), (k) and (Ω), and this implies that the spatial distribution of energy deposition events along the particle track and the geometry of the DNA helix play an important role in defining the number of double strand breaks induced in this mode of radiation action. The geometry of the DNA helix is thought to be essentially constant but the geometry of the particle track is strongly dependent on the type and energy of the incident radiation. Thus, it may be anticipated that the induction of DNA double strand breaks in proportion to dose will show a dependence on radiation type and energy, that is, a radiation quality or RBE effect.

Clearly, a 1.0 MeV proton particle track with energy deposition events occurring every 1.2 nm, that is, the separation of the two strands of the double helix, will be rather efficient in inducing a double strand break if it passes close to the DNA. On the other hand, an energetic electron particle track with scattered energy deposition events occurring every 10 nm will be rather unlikely to cause a double strand break even if it passes close to the DNA. At first sight, this might be taken to suggest that some types of radiation, such as sparsely ionising radiation, in the form of gamma-rays or X-rays, which deposit energy through fast electron tracks, would not induce double strand breaks via the alpha mode of radiation action, (α) would be zero, and at low doses this radiation would not cause an effect. However, this overlooks the fact that, as the fast electrons slow down by scattering, they eventually become low energy electrons that do have energy deposition events on the nanometre scale and are efficient at inducing DNA double strand breaks via the alpha mode. This leads to the crucially important inference that all types of ionising radiation are able to induce DNA double strand breaks at low doses down to zero dose.

Moreover, because the proportion of energy deposited by low energy electrons, compared to the total amount of energy deposited by all the energetic electrons generated by gamma or X-rays, increases as the energy of the gamma or X-rays decreases, the induction of DNA double strand breaks in the alpha mode by high energy gamma rays will be less efficient than softer gamma rays and X-rays. This means that the effect per unit dose of different sparsely ionising radiations will not all be the same at low doses.

1.8.2 THE BETA MODE OF RADIATION ACTION

$$\beta = f_0 f_1 N_2 = f_0 f_1 n\mu k (1 - \Omega k) n_1 \mu_1 k_1 D^2 \tag{1.10}$$

In Section 1.6, the parameter (f_1) was introduced into the (β) coefficient to take account of the possible repair of 'first' single strand breaks before they were converted by a 'second', independently induced break, into a double strand break. In an acute exposure, there will be little time for the repair of the 'first' strand breaks, (f_1) will be close to unity and (β) will be close to (β_∞s), but as the exposure time is increased more 'first' breaks will be repaired, (f_1) and consequently (β) will decrease, until at a sufficiently protracted exposure all the 'first' single strand breaks will be repaired, (f_1) and the (β) coefficient will become zero. The (α) coefficient which has no time dependence remains constant as the exposure is protracted so that at a very low-dose rate the number (N) of double strand breaks induced will be:

$$N = \alpha D = 2f_0 n\mu k \Omega k D. \tag{1.11}$$

This implies that at very low-dose rate the induction of DNA double strand breaks will be linear with dose from zero dose up. There is no threshold dose below which no biological effect is found.

The detailed derivation of the (β) term takes account of a series of parameters which influence the probability that a DNA double strand break is induced by two independent particle tracks. For instance, when sparsely ionising radiation (X-rays, gamma rays, fast electrons) is used to induce double strand breaks, the product (Ωk), which represents the probability per 'first' strand break that the 'second' strand is broken in the passage of the same ionising particle leading to a double strand break, will be small compared to 1, even though it may vary considerably. This means that the value of $(1 - \Omega k)$, the probability per 'first' strand break that the 'second' strand is not broken in the passage of the same ionising particle, will be close to 1. The inference is that the (β) coefficient for different sparsely ionising radiations will be almost the same when exposures are made under the same experimental conditions and at the same dose rate.

On the other hand, when the ionising radiation is 'densely ionising', such as the 1 MeV proton with energy depositions every 1.2 nm, the product (Ωk) will be close to 1, and this means that if the 'first' DNA strand is broken, the 'second' strand will also be broken by the same ionising track. The value of $(1 - \Omega k)$ will be close to zero, the quadratic (β) coefficient will be close to zero and the number (N) of double strand breaks, even for an acute exposure, will be close to:

$$N = \alpha D = 2f_0 n \mu k \Omega k D. \tag{1.12}$$

This means that the dose relationship for 'densely ionising' radiations will be dominated by the linear (α) coefficient, be considerably larger than the linear coefficient for 'sparsely ionising' radiations, and be linear with radiation dose from zero dose up.

The (β) coefficient also includes the probability $(n_1 \mu_1 k_1)$ per unit dose per 'first' single strand break that a 'second' break causing a double strand break, is induced in the passage of a separate ionising particle. A distinction has been made between a 'first break' and a 'second break'. This distinction is made because it is not known, *a priori*, that the same mechanisms are involved in the induction of the 'first break' and 'second break'. Two different radicals might be involved in the 'first break' and 'second break' or perhaps the same radical is involved but the diffusion distance from the site of induction to the site of the break is different. The inference is that the effect of exposures in different conditions, such as aerobic and anaerobic, might have different modifying effects on the (α) and (β) coefficients with probably a greater effect on the (β) coefficient.

The importance of these implications will become apparent later when the induction of double strand breaks is related to cellular biological effects and the induction of cancer and hereditary effects.

1.9   CONCLUSIONS

A detailed, parameterised linear–quadratic dose–effect equation for the induction of DNA double strand breaks by ionising radiation has been derived from first principles using the known structure and properties of the DNA molecule in eukaryotic cells. Consideration of the different parameters and coefficients leads to predictions that the quadratic (β) coefficient will be reduced, as exposures are protracted, and that the linear (α) coefficient will be strongly dependent on the spatial distribution of the energy deposition events along particle tracks, which implies that different radiations will have different biological effectiveness.

Experimental measurements of the induction of DNA double strand breaks as a function of radiation dose are presented to illustrate the linear–quadratic nature of dose–response curves in agreement with the theoretically derived equation.

2 The Molecular Model for Cellular Effects

The linear–quadratic dose–effect relationship for DNA double strand breaks is extended to give equations for cell survival, chromosome aberrations and mutations per surviving cell by relating the double strand break lesion to each of the three cellular effects. Experimental data is presented illustrating the linear–quadratic nature of the dose effects for each cellular effect. It is shown that cell survival dose–response curves measured in synchronous cells at different stages of the cell cycle can all be analysed using the linear–quadratic equation. A consistent variation of the linear coefficients and the quadratic coefficients derived from the analyses is found through the cell cycle. A dose–effect equation for the induction of mutations per irradiated cell is derived by combining the equation for mutation induction per surviving cell with the equation for cell survival. The 'micro-dosimetry' problem associated with the quadratic coefficient and the probability for two independently induced single strand breaks to be close enough together to give a double strand break at radiobiologically relevant doses is addressed. It is concluded that the 'first' single strand break is induced close to the DNA strand which renders the complementary undamaged DNA strand vulnerable to attack by a radical species induced some nanometres distant from the strand.

2.1 THE RELATIONSHIP OF DOUBLE STRAND BREAKS TO CELLULAR EFFECTS

Figure 2.1 presents, schematically, the way in which the model makes direct associations between a DNA double strand break and the three biological endpoints: cell inactivation, or rather, its inverse cell survival (S), chromosomal aberration yield (Y) and mutation frequency (M), which are all amenable to experimental measurement and quantification.

2.2 CELL SURVIVAL

Currently, it is widely assumed that a DNA double strand break is a critical lesion which is responsible for cellular killing. Cell survival (S) is, therefore, related to the number of double strand breaks per cell (N) by assuming that the loss of a cell's ability to divide and reproduce itself is directly related to the number of unrepaired DNA double strand breaks induced in the nucleus of the cell.

Thus, if $N = \alpha D + \beta D^2$ is the average number of DNA double strand breaks per cell that are induced and effective in a uniform population of synchronised cells

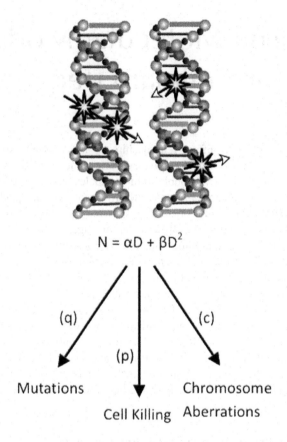

FIGURE 2.1 A schematic representation of the way in which the DNA double strand break is associated with the three cellular radiation effects, cell inactivation, chromosome aberrations and mutations.

by a dose (D) of ionising radiation, and if (p) is the probability that a double strand break leads to cell reproductive death, where (p) has a value between zero and one, so that (pN) is the average number of 'cell inactivation events' per cell, then, from the Poisson distribution of cell inactivation events per cell over the cell population, the probability for no effect, that is, cell survival (S), is given by

$$S = \exp(-pN) = \exp\left(-p\left(\alpha D + \beta D^2\right)\right). \tag{2.1}$$

In this mathematical notation, the 'cell inactivation event' can be envisaged as a 'lethal mutation'. Equation 2.1 indicates that cell inactivation results from the influence of one double strand break which causes the cell to stop dividing and does not result from the combined action of several double strand breaks in the cell.

Equation 2.1 provides a method for the quantitative analysis of single cell survival data by deriving the best fit values of the linear (pα) coefficient and the quadratic (pβ) coefficient using 'least mean squares' statistics (Chadwick and Leenhouts 1981).

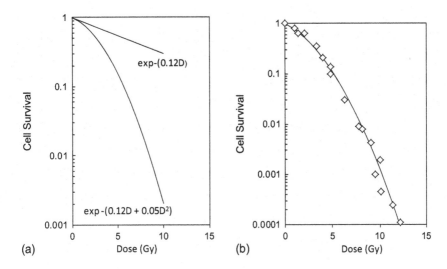

FIGURE 2.2 (a) Cell survival (S) as a function of radiation dose (D) according to Equation 2.1, broken down into its linear and quadratic components. Note the logarithmic scale for cell survival. (b) Some cell survival data plotted in the same way for comparison (data from Metting et al. 1985).

Figure 2.2b demonstrates that the equation has considerable merit for the analysis of cell survival.

A similar relationship between cell survival and radiation dose as that shown in Equation 2.1 has also been developed by Kellerer and Rossi (1971) based on a completely different approach to radiation action, namely, *dual radiation action* (Kellerer and Rossi 1972, 1978).

2.3 CHROMOSOMAL ABERRATIONS

The model proposes that, on the basis of the unineme structure of nuclear chromosomes where a single DNA double helix forms the backbone of a chromosome, a DNA double strand break is, in fact, a chromosome arm break. Therefore, the yield of chromosomal aberrations per cell (Y) is related to the number of double strand breaks per cell (N) by assuming that there is a probability (c) that each induced double strand break can lead to a detectable chromosomal aberration. Chromosomal aberrations are normally determined at the first cell mitosis following exposure and in this case there is no cause to apply Poisson statistics.

Thus, if $N = \alpha D + \beta D^2$ is the average number of DNA double strand breaks per cell that are induced and effective in a uniform population of synchronised cells by a dose (D) of ionising radiation and, if (c) is the probability that a DNA double strand break leads to a detectable chromosomal aberration, then the yield of chromosomal aberrations per cell (Y) is given by

$$Y = c(N) = c\left(\alpha D + \beta D^2\right). \tag{2.2}$$

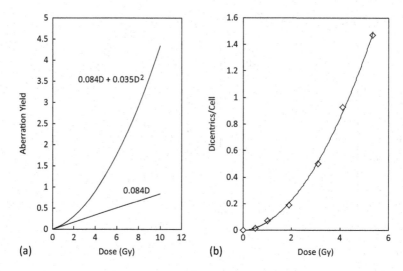

FIGURE 2.3 (a) The yield of chromosomal aberrations per cell (Y) as a function of radiation dose (D) broken down into its linear and quadratic components. (b) Some chromosomal aberration yield data for comparison (data from Lloyd et al. 1984).

Equation 2.2 provides a method for the quantitative analysis of chromosomal aberration data by deriving the best fit values of the linear ($c\alpha$) coefficient and the quadratic ($c\beta$) coefficient. Not unexpectedly, Figure 2.3b demonstrates that the equation has considerable merit for the analysis of the yield of chromosomal aberrations.

Of course, the yield of chromosomal aberrations has been analysed using linear–quadratic dose kinetics since Lea and Catcheside (1942) quantified the Classical Theory developed by Sax (1939, 1940, 1941) but the approach taken here is different. The Classical Theory, developed prior to knowledge of the double helix model of the DNA molecule and the unineme model of the chromosome, assumed that each radiation-induced chromosome break was induced in proportion to radiation dose and that a combination of two breaks to form an exchange aberration would, therefore, be linear–quadratic with dose. In the model proposed here, each chromosome break is assumed to arise from a DNA double strand break and, therefore, the dose–effect relationship for chromosome breaks will have linear–quadratic dose kinetics. In Chapter 4, the way in which different chromosomal aberrations can arise from a single chromosome break will be explained.

2.4 MUTATION FREQUENCY

The DNA double strand break is a critical lesion which disrupts the mechanical and genetic integrity of a cell and, as the perfect repair of the double strand break cannot be guaranteed, it is a potentially mutagenic lesion. The frequency of a specific mutation per cell (M) is, therefore, related to the number of double strand breaks per cell (N) by assuming that each induced double strand break has a probability (q) of inducing a specific mutation in the cell. A specific mutation is only revealed after the cell has undergone one or more cellular divisions and Poisson statistics is therefore invoked to give the mathematical relationship.

Thus, when the mutation frequency is scored per surviving cell, as is the case when a cell culture is irradiated, if $N = \alpha D + \beta D^2$ is the average number of DNA double strand breaks per cell that are induced and effective in a uniform population of synchronised cells by a dose (D) of ionising radiation, and if (q) is the probability that a DNA double strand break leads to a specific mutation, from the Poisson distribution of double strand breaks per cell over the cell population, the probability for mutation induction per surviving cell (M) is given by

$$M = \left(1 - \exp\left(-q(N)\right)\right) = \left(1 - \exp\left(-q\left(\alpha D + \beta D^2\right)\right)\right). \qquad (2.3)$$

Equation 2.3 describes a curve which shows a linear–quadratic bend upwards similar to that described by Equation 2.2 for chromosomal aberrations. However, the curve described by Equation 2.3 will eventually saturate at the value 1 when each surviving cell is mutated. This is not normally found in practice and the equation can be simplified at low doses to one comparable with Equation 2.2, that is,

$$M \approx q\left(\alpha D + \beta D^2\right). \qquad (2.4)$$

Equation 2.4 only deviates from Equation 2.3 at mutation frequencies above 0.25 mutations per cell and, as this is rarely achieved in experiments, Equation 2.4 can be used as a simplified proxy for Equation 2.3.

However, it is important to realise, for instance, that when an organism is irradiated, a cell must survive to express the mutation and this means that the mathematical formulation for mutation frequency per irradiated cell must contain a term for the induction of the mutation and a term for the survival of the cell.

Thus, the probability for mutation induction per irradiated cell (M_1) is given by

$$M_1 = M.S = \left(1 - \exp\left(-q\left(\alpha D + \beta D^2\right)\right)\right)\exp\left(-p\left(\alpha D + \beta D^2\right)\right). \qquad (2.5)$$

Equation 2.5 describes a curve which shows a linear–quadratic bend upwards from the origin, passes through a maximum mutation frequency and decreases at higher doses because of increased cell inactivation. The peak mutation frequency is related to the values of the probabilities (p) and (q) which relate DNA double strand breaks to cell inactivation and the specific mutation, respectively.

In Figure 2.5a, curve $(1) = (1 - \exp(-(0.002D + 0.003D^2))) \cdot \exp(-6(0.002D + 0.003D^2))$, and curve $(2) = (1 - \exp(0.002D)) \cdot \exp(-6(0.002D))$.

Equations 2.3 and 2.4 each provide a method for the quantitative analysis of mutation frequency per surviving cell data by deriving the best fit values of the linear $(q\alpha)$ coefficient and the quadratic $(q\beta)$ coefficient. Equation 2.5 provides a method for the quantitative analysis of mutation frequency per irradiated cell data by deriving the best fit values of the linear $(q\alpha)$ and $(p\alpha)$ coefficients and the quadratic $(q\beta)$ and $(p\beta)$ coefficients. Figures 2.4b and 2.5b demonstrate that the relevant equations have considerable merit for the analysis of the mutation frequency induced in irradiated cells.

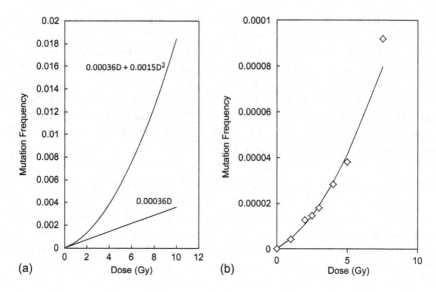

FIGURE 2.4   (a) The mutation frequency per surviving cell (M) as a function of radiation dose (D) broken down into its linear and quadratic components. (b) Some data for comparison (data from Cox et al. 1977).

The schematic representation of the model shown in Figure 2.1 suggests that the inferences, implications and insights derived for the two modes of radiation action in Section 1.8 will apply in equal measure to each cellular endpoint, survival, chromosome aberration yield and mutation frequency. Thus, it can be concluded that the cellular effects of all types of ionising radiation will be induced from zero dose up, will be proportional with dose at low doses and low-dose rates, and different types of radiation will have a different effectiveness per unit dose.

2.5   THE LINEAR–QUADRATIC EQUATION AND DATA ANALYSIS

2.5.1   SURVIVAL OF SYNCHRONOUS CELLS

The equation for cell survival,

$$S = \exp(-pN) = \exp\left(-p\left(\alpha D + \beta D^2\right)\right), \qquad (2.1)$$

is, by definition, applicable to a uniform population of single cells which is achieved, in practice, in cells synchronised in one part of the cell cycle. The mitotic cell cycle is, by convention, divided up into G1 (gap 1) when the cell prepares for DNA synthesis (S phase) and the chromosomes are replicated. A second gap phase (G2) follows when the cell prepares for mitosis (M) when the chromosomes condense and the cell divides (Chadwick and Leenhouts 1981).

Data available on the survival of synchronised cells reveal that the shape of the survival curve changes considerably as the cells move through the cell cycle. Analysis of the survival data (Chadwick and Leenhouts 1973b, 1975) revealed that Equation 2.1

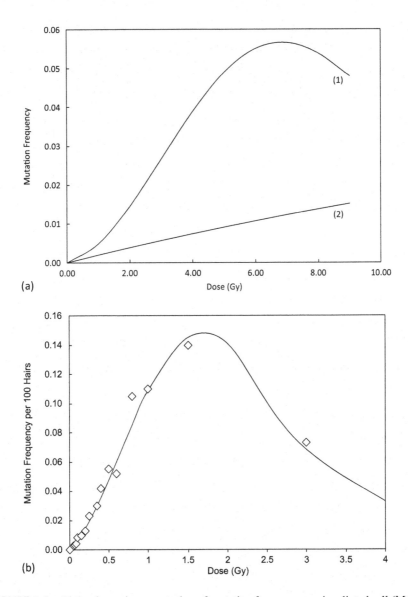

FIGURE 2.5 (a) A schematic presentation of mutation frequency per irradiated cell (M_I) as a function of radiation dose (D) broken down into its linear–quadratic and linear components. (b) Some data on mutations induced in *Tradescantia* stamen hairs presented for comparison (data from Underbrink et al. 1975).

can be accurately fitted to the synchronized cell survival in all parts of the cell cycle. Indeed, a careful study by Gillespie and his colleagues, Chapman, Reuvers and Dugle (Gillespie et al. 1975a,b), showed that the equation fitted their experimental data as well as could be expected on statistical grounds (see Figure 2.6).

A consistent variation in the values of the linear (pα) and quadratic (pβ) coefficients through the cell cycle has been found in the analysis of a series of different

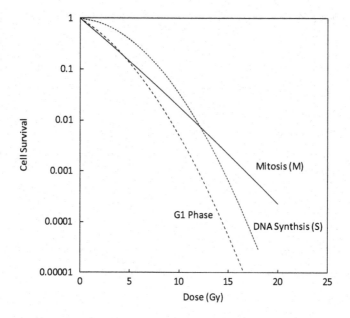

FIGURE 2.6 The survival of cells synchronised in different phases of the cell cycle illustrating the changing shape of the survival curve (drawn after Kruuv and Sinclair 1966). The shape changes from very linear when the cells are in mitosis to very quadratic when the cells are in the S or DNA synthesis phase.

synchronised cells (Chadwick and Leenhouts 1975). The linear (pα) coefficient exhibits a substantial reduction as the cells pass through the DNA synthesis phase of the cell cycle and the quadratic (pβ) coefficient exhibits a peak value at the beginning of DNA synthesis and the lowest value in mitosis (see Figure 2.7b).

Interestingly, the yield of chromosomal aberrations in synchronised cells has also been shown to vary as these cells pass through the cell cycle and, analysis of the dose–effect curves using the equation

$$Y = cN = c\left(\alpha D + \beta D^2\right) \tag{2.2}$$

reveals a comparable variation in the linear and quadratic coefficients as that found for cell survival (see Figure 2.8).

This is not surprising, as Figure 2.1, showing the link from the DNA double strand break lesion to the different cellular endpoints, implies that the different endpoints should show comparable responses under different irradiation conditions.

Analysis, in terms of the variation of the induction of DNA double strand breaks, permits some speculation that the dip in the linear (pα) coefficient in the DNA synthesis phase of the cell cycle might reflect a loosening of the DNA as replication starts (Benbow et al. 1985; Linskens and Huberman 1990; Douglas et al. 2018). Consequently, the probability (Ωk) of a single radiation track passing close to both DNA strands and causing breaks in each strand will decrease. Some support for this speculation can be found in the changing magnitude of the dip when more densely ionising radiations

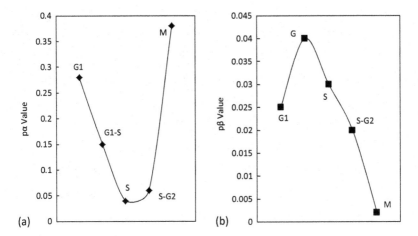

FIGURE 2.7 (a) An example of the consistent variation found in the value of the linear coefficient (pα) of synchronised cells through the cell cycle. (b) An example of the consistent variation found in the value of the quadratic coefficient (pβ) of synchronised cells through the cell cycle.

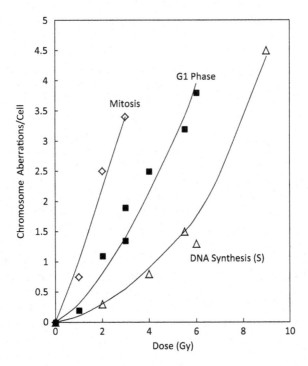

FIGURE 2.8 The changing shape of the yield of chromosome aberrations in synchronised cells as the cells progress through the cell cycle. This imitates that found for survival (data from Dewey et al. 1970, 1971a).

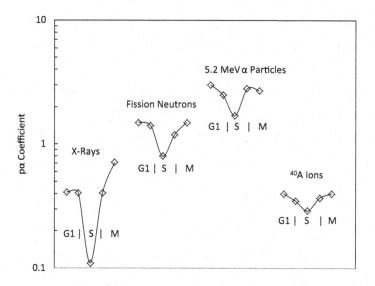

FIGURE 2.9   The variation in the linear (pα) coefficient for synchronised cells as the cells pass through the cell cycle following exposure to different types of radiation. The dip in the DNA synthesis (S) phase decreases as the radiation becomes more densely ionising (data from Sinclair and Morton 1966; Sinclair 1969; Hall et al. 1972; Bird and Burki 1971).

are used. Figure 2.9 shows this for gamma rays, neutrons, alpha particles and argon ions and reveals that although the value of (pα) increases with more densely ionising radiation, the dip is less exaggerated as the radiation becomes more densely ionising. Indeed, the dip is hardly significant for the very densely ionising argon ions. The reason that the value of the linear (pα) coefficient for the argon ions is much lower than the value for alpha particles, even though the argon ions are more densely ionising than the alpha particles, is that the argon ions deposit much more energy in the DNA than is needed simply to create the double strand break. In other words, there is 'overkill'.

The peak value in the quadratic (pβ) coefficient occurring at the beginning of the DNA replication phase of the cell cycle could also be related to a loosening of the DNA strands which allows easier radical access. In this respect, the almost purely linear survival of mitotic cells when the DNA is tightly wrapped in the condensed chromosomes is potentially significant. It would appear that the 'packaging' of the DNA in the cell nucleus affects the balance between the production of double strand breaks in the alpha mode and in the beta mode as the cell goes through the cell cycle.

Gillespie and his colleagues (Gillespie et al. 1975a,b) also investigated whether the linear–quadratic Equation 2.1 could be used to analyse an asynchronous population of cells. They used appropriately weighted data from synchronised cells to create a cell survival curve for an exponentially growing cell population and successfully reanalysed the curve using the linear–quadratic Equation 2.1. They also took data from four different experiments measuring survival in exponentially growing cells to synthesise a survival curve and found that the synthesised curve could also be successfully fitted using Equation 2.1. Gillespie and his colleagues rightly indicate that the values of the linear and quadratic coefficients, determined in an analysis

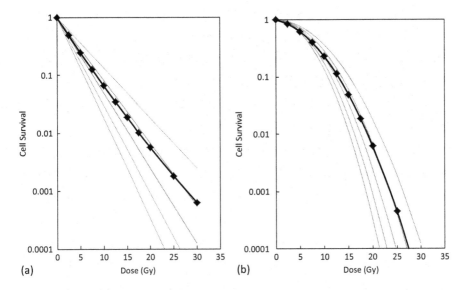

FIGURE 2.10 (a) The solid line denotes the survival which would be found in an asynchronous population of cells made up in equal proportions of cells having the five linear exponential survival curves illustrated by the dotted lines. The solid line has a negative quadratic coefficient and tails out. (b) The solid line denotes the survival which would be found in an asynchronous population of cells made up in equal proportions of cells having the five purely quadratic exponential survival curves illustrated by the dotted lines. The solid line has a positive initial linear coefficient.

of an asynchronous population of cells, need to be interpreted with some caution. This is illustrated in Figure 2.10a, where a combination of purely linear survival curves, when mixed, leads to a curve tailing out which gives a negative value for the quadratic coefficient. When a mixed curve of purely quadratic survival curves is created and analysed this gives a survival curve with a positive linear coefficient (see Figure 2.10b). Gillespie et al. (1975a,b) have aptly called this 'crosstalk' between the linear and quadratic coefficients.

This means that, in an analysis of cell survival in an asynchronous population, the quadratic coefficient will be slightly weakened and the linear coefficient will be slightly strengthened.

It can be concluded that Equation 2.1, although defined for a uniform population of cells, can also be applied to the analysis of survival of an asynchronous population of cells and that the same reasoning can be applied to measurements of chromosomal aberrations and mutation frequency.

2.5.2 The Influence of Different Exposure Conditions on the Linear and Quadratic Coefficients

Another result from the use of the linear–quadratic equation for the analysis of cell survival curves is the indication that changes in the shape of the curves, as a consequence of the exposure of cells under different conditions, such as aerobic and

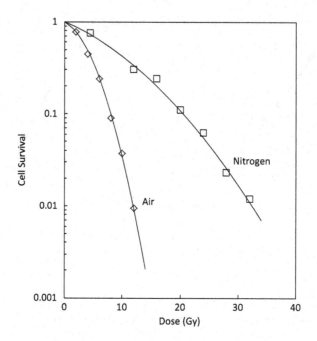

FIGURE 2.11 The survival of Chinese hamster cells in air and in nitrogen. Fitting of Equation 2.1 gives for air, $(p\alpha) = 0.077Gy^{-1}$ and $(p\beta) = 0.026Gy^{-2}$; and for nitrogen, $(p\alpha) = 0.061Gy^{-1}$ and $(p\beta) = 0.0025Gy^{-2}$ (data from Chapman et al. 1975a).

hypoxic, lead to a larger relative change in the quadratic $(p\beta)$ coefficient compared with the relative change in the linear $(p\alpha)$ coefficient. This can be seen in Figures 2.11 and 2.12 which illustrate the influence of changes in the oxygenation of the cells on the radiation effect (Chapman et al. 1975a; Cooke et al. 1976).

In the case of Chapman et al. (1975a), the ratio of the air to nitrogen linear $(p\alpha)$ coefficients is approximately 1.25 and the ratio of the air to nitrogen quadratic $(p\beta)$ coefficients is approximately 10.

The analysis of the Cooke et al. (1976) data gives a ratio of aerobic to anaerobic linear $(p\alpha)$ coefficients of approximately 2.8 and the ratio of the aerobic to anaerobic quadratic $(p\beta)$ coefficients of approximately 8.

The enhanced effect of oxygen can also be seen in Figure 1.9 for the induction of DNA double strand breaks in aerobic and hypoxic cells (Prise et al. 1987) where the ratio of the aerobic to hypoxic linear (α) coefficients is approximately 4.5 and the ratio of the aerobic to hypoxic quadratic (β) coefficients is approximately 7.35. The effect of oxygen has almost double the effect on the quadratic coefficient as on the linear coefficient.

This can be quantified quite well by making use of the expressions for the linear and quadratic coefficients derived in Chapter 1.

Equation 1.8 gives the number of induced and effective double strand breaks as:

$$N = \left(\alpha D + \beta D^2\right) = 2f_0 n\mu k\Omega kD + f_0 f_1 n\mu k\ (1 - \Omega k)\,n_1\,\mu_1 k_1 D^2.$$

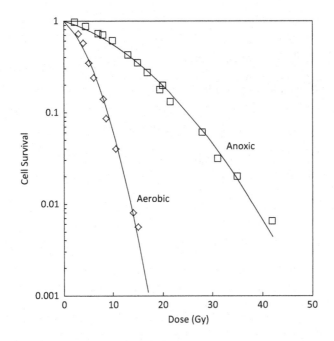

FIGURE 2.12 The survival of Chinese hamster cells under aerobic and anoxic conditions. Fitting of Equation 2.1 gives for aerobic conditions, $(p\alpha) = 0.107Gy^{-1}$ and $(p\beta) = 0.0173Gy^{-2}$; and for anoxic conditions, $(p\alpha) = 0.038Gy^{-1}$ and $(p\beta) = 0.00217Gy^{-2}$ (data from Cooke et al. 1976).

If $\alpha_A = 2f_0n\mu k_A\Omega k_A$ represents the linear coefficient in air and $\alpha_B = 2f_0n\mu k_B\Omega k_B$ represents the linear coefficient in hypoxia, where it is assumed that the influence of the different exposure conditions only affects the quantity (k), which is dependent on the chemical environment around the DNA, and more specifically on the radical scavenging, in the nucleus of the cell.

Then, the ratio of the linear coefficients becomes:

$$\alpha_A/\alpha_B = 2f_0n\mu k_A\Omega k_A/2f_0n\mu k_B\Omega k_B = k_A^2/k_B^2. \qquad (2.6)$$

This ratio gives a measure of the oxygen effect on the linear coefficient.

If $\beta_A = f_0f_1n\mu k_A(1-\Omega k_A)n_1\mu_1 k_{1A}$ represents the quadratic coefficient in air and $\beta_B = f_0f_1n\mu k_B(1-\Omega k_B)n_1\mu_1 k_{1B}$ represents the quadratic coefficient in hypoxia, where it is assumed that the influence of the different exposure conditions only affect the quantities (k) and (k_1), which are dependent on the chemical environment around the DNA, and more specifically on the radical scavenging, in the nucleus of the cell.

Then, the ratio of the quadratic coefficients becomes:

$$\beta_A/\beta_B = f_0f_1n\mu k_A\left(1-\Omega k_A\right)n_1\mu_1 k_{1A}/f_0f_1n\mu k_B\left(1-\Omega k_B\right)n_1\mu_1 k_{1B},$$

that is,

$$k_A\left(1-\Omega k_A\right)k_{1A} / k_B\left(1-\Omega k_B\right)k_{1B}$$

but, as (Ωk_A) and (Ωk_B), the probabilities per 'first' strand break that the 'second' strand is broken in the passage of the same ionising particle, are both very much less than 1 for sparsely ionising radiation, this equation can be approximated to:

$$\beta_A / \beta_B = k_A k_{1A} / k_B k_{1B} = \left(\alpha_A / \alpha_B\right)^{1/2} k_{1A} / k_{1B}. \qquad (2.7)$$

The data of Chapman et al. (1975a) give $(\alpha_A/\alpha_B)^{1/2} = 1.118$ and $(k_{1A}/k_{1B}) = 8.94$. The data of Cooke et al. (1976) give $(\alpha_A/\alpha_B)^{1/2} = 1.67$ and $(k_{1A}/k_{1B}) = 4.8$. The data of Prise et al. (1987) give $(\alpha_A/\alpha_B)^{1/2} = 2.12$ and $(k_{1A}/k_{1B}) = 3.46$.

All these results indicate a larger oxygen effect on the quadratic (β) coefficient compared with the linear (α) coefficient. The important conclusion to be drawn from this is that a different process is involved in the 'second' single strand break of the quadratic (β) coefficient compared with the processes involved in both the 'first' single strand break as well as the two simultaneous single strand breaks of the linear (α) coefficient. Thus, the distinction that has been made between the parameter (k) involved in the induction of the 'first' breaks and the parameter (k_1) involved in the induction of the 'second' breaks is justified. However, this means that Figure 1.3, widely used to establish the basis for the derivation of the linear–quadratic dose–effect relationship for the induction of DNA double strand breaks, is somewhat misleading and should be modified. And this means that a reconsideration of the process leading to the 'second' single strand break in the quadratic (β) coefficient is required.

2.6   THE MICRO-DOSIMETRY PROBLEM

The micro-dosimetry problem arises from a radiation physics consideration of the derivation of the linear–quadratic dose–effect function for the induction of DNA double strand breaks and, more specifically, the derivation of the quadratic (β) coefficient in the beta mode. The reason for this is that a calculation of the probability of two independent ionising radiation tracks occurring within two nanometres, that is, opposite, or even closely opposite, to each other on the two DNA strands to create the two independent single strand breaks, indicates that very large doses are needed before the probability becomes significant (Kellerer 1975). In other words, the calculation implies that at biologically significant doses the quadratic term (βD^2) would be minute and the dose–effect relationship for the induction of DNA double strand breaks would be purely linear with dose. This has been a major problem for the molecular model, outlined here, and has certainly impeded its wider acceptance and application. To some extent the use of Figure 1.3 as an illustration for the derivation of the linear–quadratic dose–effect relation for DNA double strand breaks, showing the two independent ionisation events on the DNA strand in the beta mode, has probably contributed to the problem.

In fact, such a calculation of the quadratic coefficient is primarily dependent on radiation physics and on the dimensions chosen for the interaction events. It assumes that the same interaction process between the radiation and the DNA strands occurs for both the 'first' and the 'second' independent single strand breaks. It neglects any influence that the occurrence of the 'first' single strand break may have on the chemistry, biochemistry, distortion and 'packaging' of the DNA helix which might render the DNA vulnerable to a different interaction process creating the 'second' single strand break.

The results presented in the previous section and Figures 2.11 and 2.12 on the greater effect of oxygen on the quadratic (β) coefficient compared to the linear (α) coefficient indicate the need to take account of two different processes being involved in the 'first' single strand break and the independently induced 'second' break. The wide variation in the shape of the survival curve of cells synchronised in different phases of the cell cycle illustrates the influence of alterations in the 'packaging' and biochemical surroundings of the DNA double helix as the cells pass through the cell cycle. Indeed, the situation in mitosis, when the cell survival curves have a very small quadratic (β) coefficient and the DNA is wrapped into the condensed chromosomes, might approximate to the expectation of the 'micro-dosimetric' calculations, although that does not apply to other phases of the cell cycle.

In addition, there is now an abundance of experimental evidence documenting the linear–quadratic dose–effect relationship for the induction of DNA double strand breaks in cells at doses of radiobiological significance in a wide variety of different exposure conditions, as illustrated in Figures 1.5 to 1.13.

It seems reasonable, therefore, to accept the validity of the linear–quadratic dose–effect relationship for the induction of DNA double strand breaks and cellular radiation effects, in general, but to acknowledge that a reassessment of the processes involved in the quadratic (β) coefficient deserves further consideration.

Chapman and his colleagues (Chapman et al. 1975b, 1976) have investigated the changes in the linear ($p\alpha$) coefficient and quadratic ($p\beta$) coefficient of cell survival using the concentration dependence of radical scavenging to alter the conditions of the cells during exposure. They defined two average diffusion distances for the radiolysis products and concluded that the linear ($p\alpha$) coefficient arises from OH radicals diffusing some 0.8 nm to attack the DNA helix, while the distance involved in the quadratic ($p\beta$) coefficient was some 2 to 3 nm.

Using a relatively simple analogue track model, but one which allowed the spatial distribution of ionisation events in relation to the spatial dimensions of the DNA helix to be taken into account, it was shown that experimental data on cell inactivation in *Chlorella* cells made as a function of the stopping power of different radiations (Roux 1974) could be adequately explained by proposing that the linear ($p\alpha$) coefficient arose as a result of the occurrence of two radicals very close (0.35–0.7 nm) to the DNA helix. The quadratic ($p\beta$) coefficient could be explained by the occurrence of a 'first' single strand break caused by a radical occurring very close (0.35–0.7 nm) to the DNA helix combined with a 'secondary' event caused by a different active species induced up to some 8.5 nm from the DNA helix (Leenhouts and Chadwick 1976; Chadwick and Leenhouts 1981).

The micro-dosimetry problem with the quadratic coefficient ($p\beta$) devolves, at this stage, into a process which combines a single strand break induced close to the DNA helix and a subsequent process that permits a 'second' energy deposition event from an independent radiation track, some distance from the DNA helix, to convert the single strand break into a double strand break. In other words, the initial single strand break creates a vulnerable target on the remaining intact DNA strand which is broken in the 'second' event. It is known that the DNA helix is under some tension (Bentham 1979; Ljungman and Hanawalt 1992) and might be expected to unravel a little when one strand is broken and it is also known that DNA that is not protected by, for example, histones, is increasingly sensitive to radical attack (Ljungman et al. 1991). These factors may play a role in this process. In addition, the 'loosening' of the DNA helix as the cell enters DNA synthesis and the occurrence of 'micro-bubbles' of open DNA during DNA replication (Benbow et al. 1985; Gaudette and Benbow 1986) might contribute to a peak in the quadratic ($p\beta$) coefficient found in synchronised cells at the start of DNA synthesis. It also appears that a 'bubble' of open DNA occurs during DNA transcription (Hahn and Buratowski 2016; Plaschka et al. 2016, Yuan et al. 2016).

It is important to note that this 'second' event can only convert a pre-existing single strand break to a double strand break. It cannot induce a single strand break on an undamaged DNA double helix. In fact, it seems that an intact DNA helix can only be damaged by an energy-deposition-induced radiolysis product formed very close to the DNA strand (Powers 1974). Consequently, it appears probable that DNA single strand breaks are only induced by the passage of a radiation track very close (0.35–0.7 nm) to the DNA helix. Evidence in support of this can be found in the much reduced yield of single strand breaks and the reduced quadratic ($p\beta$) coefficient found after exposure to very densely ionising radiations (Christensen et al. 1972; Barendsen et al. 1966; Todd, 1967; Roux 1974). These densely ionising radiations are, nevertheless, efficient in inducing DNA double strand breaks in the passage of a single ionising particle track (see Chapter 6).

Therefore, it appears that the beta mode induction of double strand breaks shown in Figure 1.3 should be replaced by the sequence of events shown in Figure 2.13.

This figure shows, in the beta mode, the 'first' single strand break induced by an ionisation energy deposition very close (0.35–0.7 nm) to the intact DNA double helix and a 'second' ionisation energy deposition arising from a separate ionising particle track occurring some distance (8 nm) from the damage on the DNA helix which converts the 'first' single strand break into a double strand break. The 'first' single strand break would make the damaged site on the DNA helix vulnerable and possibly more sensitive to the radical created some distance from the damaged site.

The description of the beta mode of double strand break production changes in as much as the process should be seen as a 'first' single strand break rendering, very quickly, the DNA at that single strand break vulnerable to a 'second' event which converts the single strand break to a double strand break. Without the 'first' single strand break, the undamaged DNA helix would not be damaged or broken by the 'second' event. However, the mathematical derivation, outlined in Chapter 1, would not be altered because account has been taken of the possibility that a different process might be involved in the beta mode. In the equations, (k_1) is different from (k) and this accounts for the larger influence of chemical changes in the cell environment during exposure on the quadratic (β) coefficient compared with the effect on the linear (α) coefficient.

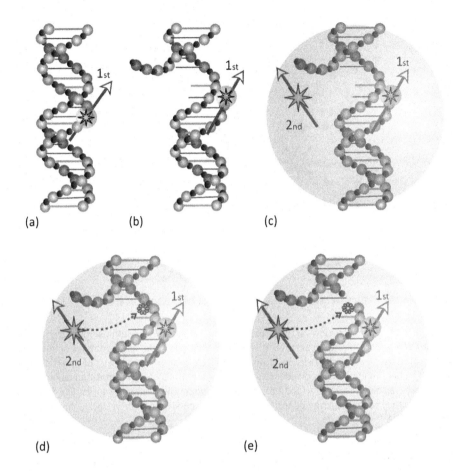

FIGURE 2.13 The sequence of events which is thought to occur, based on experimental and theoretical data analyses (Chapman et al. 1976; Leenhouts and Chadwick 1976; Chadwick and Leenhouts 1981), in the induction of a DNA double strand break by two independent radiation tracks. (a) An energy deposition very close (see very small opaque sphere) to the DNA induces a single strand break, (b) the DNA helix unravels a little, (c) a second radiation track some distance from the DNA helix (see the larger opaque sphere) creates an energy deposition, (d) an active species from that energy deposition diffuses to attack the intact but vulnerable DNA strand to cause, (e) a break which converts the 'first' single strand break into a double strand break. The opaque spheres represent the volumes within which any energy depositions can have an effect on the DNA. The 'first' sphere is very small but the 'second' sphere is much larger, in line with the diffusion distances derived in experimental and theoretical analyses.

Consequently, Figure 1.3 should be replaced by Figure 2.14.

One further consequence of the revised beta mode of double strand break induction concerns the scavenging of the linear (α) coefficient defined by (k_A^2/k_B^2) (see Equation 2.6) and the scavenging of DNA single strand breaks which would be represented by (k_A/k_B). In other words, the scavenging of DNA double strand breaks induced in the alpha mode should be the square of the scavenging of DNA single strand

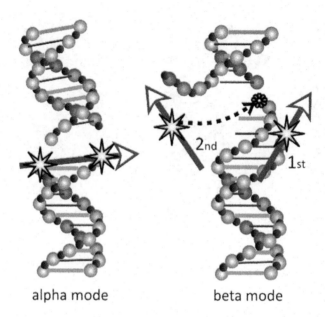

alpha mode   beta mode

FIGURE 2.14 The new view on the linear–quadratic dose–effect relationship for the induction of DNA double strand breaks by ionising radiation. The alpha mode remains unchanged and involves two energy depositions close to the DNA helix in the passage of one radiation track. The beta mode is now considered to arise from a 'first' energy deposition close to the DNA helix, causing a single strand break, which renders the remaining intact DNA strand vulnerable to a 'second' energy deposition from an independent radiation track some distance from the DNA.

breaks measured in the same experiment. The data reported by Tisljar et al. (1983) satisfy this requirement for tritium exposure with an oxygen enhancement for double strand breaks of 3.6 and an oxygen enhancement for single strand breaks of 1.7. However, it is difficult to find other data where these measurements have been made in the same experiment.

The torsional stress on the DNA and its potential unravelling as a consequence of a DNA single strand break raises the question of timing and how fast the unravelling might be. Can the unravelling time be beaten with ultra-fast exposures and does the quadratic (β) coefficient disappear or is the active radiolytic product of the 'second' radiation event long-lived and able to diffuse to the vulnerable site on the DNA opposite the 'first' single strand break either during or immediately after an ultra-fast exposure?

2.7 CONCLUSIONS

The linear–quadratic dose–effect relationship for DNA double strand breaks is extended to give equations for cell survival, chromosome aberrations and mutations per surviving cell by relating the double strand break lesion to each of the three cellular effects. Experimental data are presented illustrating the linear–quadratic nature of the dose effects for each cellular effect. Values for the linear (α) and quadratic (β) coefficients can be derived from fitting the equation to the experimental data.

The 'micro-dosimetry' problem associated with the quadratic coefficient and the probability for two independently induced single strand breaks to be close enough together to give a double strand break at radiobiologically relevant doses is addressed and resolved. It is concluded that the 'first' single strand break is induced close to the DNA strand which renders the complementary undamaged DNA strand vulnerable to attack by a radical species induced some nanometres distant from the strand.

3 The Link from Molecular Lesion to Cellular Effects

The associations made in the previous chapter between the radiation-induced DNA double strand breaks and the three cellular effects lead to a series of predicted correlations which are derived in a series of equations. Experimental data from three different laboratories are presented demonstrating the direct correlation between induced DNA double strand breaks and cell survival when measured in the same experiment. The data cover a wide range of cell types and exposure conditions. Other experimental data is presented demonstrating the direct correlation between cell survival and chromosome aberrations and also between cell survival and mutation frequency per surviving cell. All these data provide evidence of the direct link between the molecular lesion, the DNA double strand break and the three cellular effects which is predicted by the model. The different correlations measured for more densely ionising radiations in contrast to the correlation measured for sparsely ionising radiations is addressed. It is concluded that the induction of double strand break clusters by densely ionising radiations and the probably incorrect estimation of double strand break numbers is responsible for the discrepancy. It is shown that the same peak height of the mutation frequency per irradiated cell dose–effect relationship is a further correlation also corroborated by experimental analyses.

3.1 THE CORRELATIONS BETWEEN DNA DOUBLE STRAND BREAKS AND CELLULAR EFFECTS

Figure 3.1 presents the routes by which a direct linkage chain can be established from the induction of DNA double strand breaks to the cellular effects such as cell survival, chromosomal aberrations and somatic mutations. Figure 3.1 is based on Figure 2.1 but the original alpha mode - beta mode drawing of the double strand breaks has been replaced by the revised view as illustrated in Figure 2.14 and the direct routes of the linkage chain have been highlighted.

In Chapter 1, a linear–quadratic equation was developed for the induction of DNA double strand breaks and a series of experimental measurements were presented in Figures 1.5 to 1.13 demonstrating the curved nature of the dose–effect relationship for the induction of double strand breaks. In Chapter 2, the DNA double strand break was associated with cell inactivation, chromosomal aberrations and mutations and linear–quadratic dose–effect relationships were derived for the three cellular effects and were shown to provide very good fits to experimental data in Figures 2.2 to 2.5. In Figure 3.1, the association of the DNA double strand break with the three cellular effects is taken a stage further and a linkage chain is indicated from the molecular lesion to the three cellular effects.

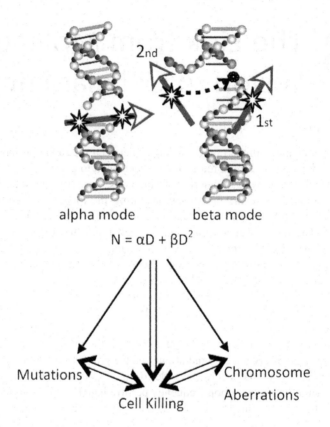

FIGURE 3.1 A schematic representation of the linkage chain from the molecular lesion, the double strand break, to the three cellular radiobiological effects, highlighting the direct link to cell inactivation and also the further linkage from cell inactivation to chromosome aberrations and mutations.

3.2   THE CORRELATION BETWEEN DNA DOUBLE STRAND BREAKS AND CELL SURVIVAL

Equation 3.1 provides a direct quantitative relationship between cell survival (S) and the number of induced and effective DNA double strand breaks (N)

$$S = \exp(-p(\alpha D + \beta D^2)) = \exp(-pN) \tag{3.1}$$

so that

$$\ln S = -pN. \tag{3.2}$$

Equation 3.2 implies that the number of DNA double strand breaks is expected to be linearly related to the logarithm of cell survival when the measurements are made on the same population of irradiated cells. The equation is independent of the values of the linear (α) coefficient and the quadratic (β) coefficient and this means that the same correlation should be found following different exposure conditions in the same cell system.

In Chapter 1, several figures were presented documenting the linear–quadratic dose–effect relationship for the induction of DNA double strand breaks. In many cases, cell survival was also measured in the same experiments and the correlation defined by Equation 3.2 was investigated. The first to do these measurements was Radford (1985, 1986) who used several different exposure conditions and also investigated three different cell systems. In 1987, Prise and his colleagues (Prise et al. 1987) published similar measurements, using aerobic and anoxic cells and also different radiations and, in 1989, Murray and his colleagues (Murray et al. 1989) published data using different radio-protectors.

The following series of figures presents these results, redrawn from the originals in the form of the dose–effects for cell survival and DNA double strand break induction, together with the correlations according to Equation 3.2.

The result, that a single correlation was found between DNA double strand breaks and cell survival for three different cell lines from humans, hamsters and mice, led Radford to state that 'for each cell type there is a comparable probability of conversion of a DNA dsb into a lethal lesion' (Radford 1986).

The data and correlations shown in Figures 3.2 through 3.7, which originate in three laboratories on three different continents and involve multiple different exposure conditions and several different cell lines, provide a compelling direct link from radiation-induced DNA double strand breaks to cell inactivation. The correlations are in accordance with Equations 3.1 and 3.2 and thus provide the first link in the chain, elaborated in the model and illustrated in Figure 3.1.

Nevertheless, it needs to be noted that all the data and correlations shown refer to sparsely ionising radiation. However, Prise et al. (1987, 1990; Prise 1994) and Folkard et al. (1989) have investigated the induction of DNA double strand breaks and cell survival using several more densely ionising radiations. They found that the shape of the dose–effect curves for induced DNA double strand breaks followed the expected change in shape, becoming less curved and eventually linear as the radiation became more densely ionising and more efficient in inducing DNA double strand breaks in the alpha mode but they also found that the correlation between the induced DNA double strand breaks and cell survival changed and the slope of the correlation increased, indicating fewer double strand breaks per lethal event (see Figure 3.8).

A problem which arises when the neutral filter elution method is used to measure DNA double strand breaks after exposure of cells to densely ionising radiation is that the densely ionising radiation creates clustered double strand breaks (Rydberg 1996, 2001). This means that the double strand breaks are not induced randomly in the cell nuclear DNA, contrary to the situation with sparsely ionising radiation. There is evidence to suggest that these clustered double strand breaks are only counted as a single double strand break (Rydberg et al. 1994) and that the actual number of double strand breaks induced by densely ionising radiation would be seriously underestimated by the neutral filter elution measurements.

In fact, Prise et al. (1989) measured DNA double strand breaks and cell survival using ultrasoft aluminium-K X-rays which will not induce DNA double strand breaks in clusters but in a more random distribution. The aluminium-K X-rays have an energy of 1.5 keV, a track length of 70 nanometres and an LET (linear energy

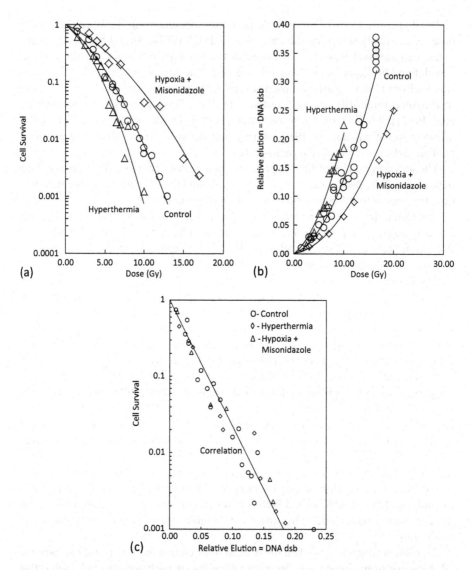

FIGURE 3.2 Data from Radford (1985) in mouse L cells exposed in air (control), hypoxia plus misonidazole and after hyperthemia. The survival curves (a) reflect the induction of DNA double strand break curves (b) (already seen as Figure 1.6). The straight line correlation (c) between survival and the number of double strand breaks, in accordance with Equation 3.2, is common for all three treatments.

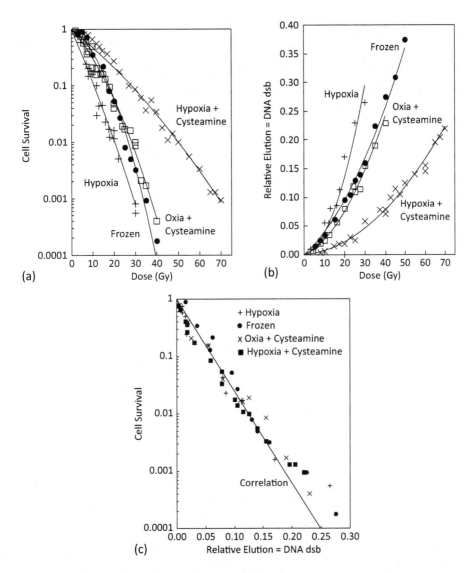

FIGURE 3.3 Data from Radford (1985) in mouse L cells exposed frozen, in hypoxia, in air with cysteamine and in hypoxia with cysteamine. The survival curves (a) reflect the induction of DNA double strand break curves (b) (already seen as Figure 1.7). The straight line correlation (c) between survival and the number of double strand breaks, in accordance with Equation 3.2, is common for all four treatments.

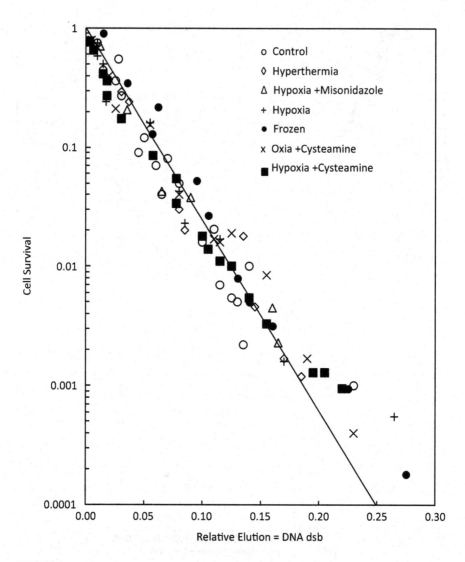

FIGURE 3.4 The single correlation found by Radford (1985) for all seven exposure conditions in accordance with Equation 3.2. The figure combines the correlations shown in Figures 3.2 and 3.3.

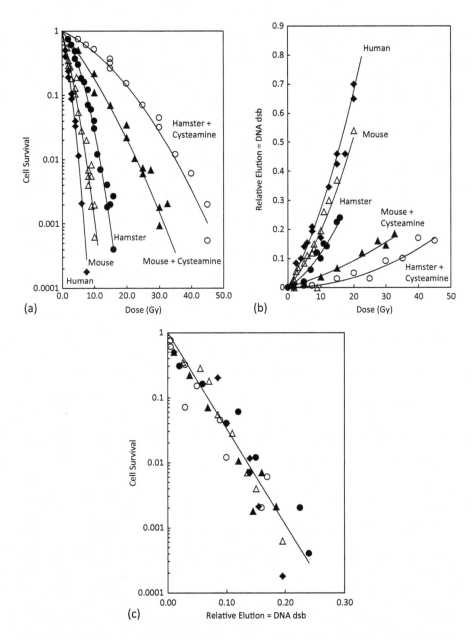

FIGURE 3.5 Data from Radford (1986) on cell survival and the induction of DNA double strand breaks in human cells (♦) exposed in air and V79 Chinese hamster (●,○) and mouse fibroblast cells (△,▲) exposed in air and with the protector cysteamine. The survival curves (a) reflect the induction of DNA double strand break curves (b). The straight line correlation (c) between survival and the number of double strand breaks, in accordance with Equation 3.2, is common for the three different cell lines.

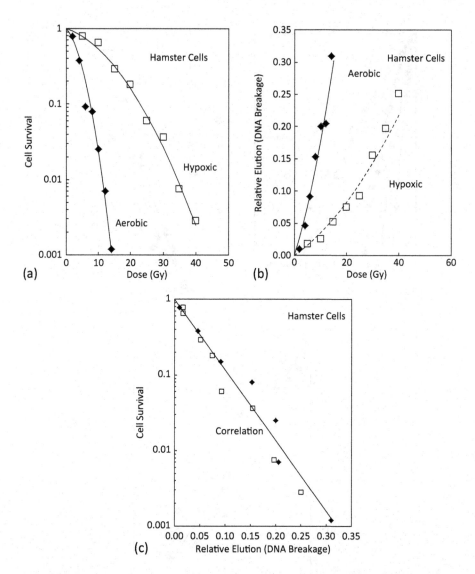

FIGURE 3.6 Data from Prise et al. (1987) on cell survival and the induction of DNA double strand breaks in V79 Chinese hamster cells exposed in air (◆) and in hypoxia (□). The survival curves (a) reflect the induction of DNA double strand break curves (b). The straight line correlation (c) between survival and the number of double strand breaks is in accordance with Equation 3.2.

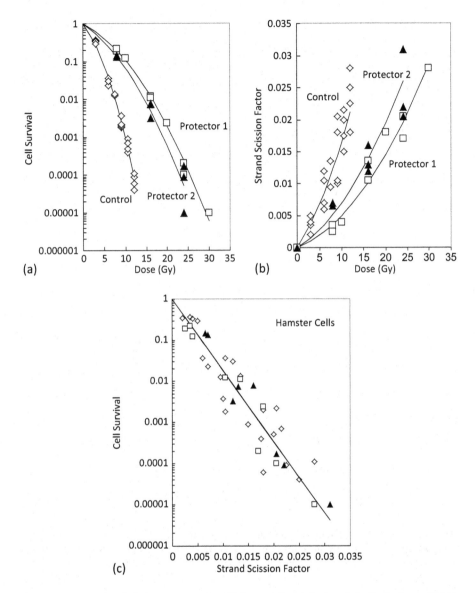

FIGURE 3.7 Data from Murray et al. (1990) on cell survival and the induction of DNA double strand breaks in Chinese hamster ovary (CHO) cells exposed to gamma rays, control (◇), protector 1 (□) and protector 2 (▲). The survival curves (a) reflect the induction of DNA double strand break curves (b) and the straight line correlation (c) between survival and the number of double strand breaks is in accordance with Equation 3.2.

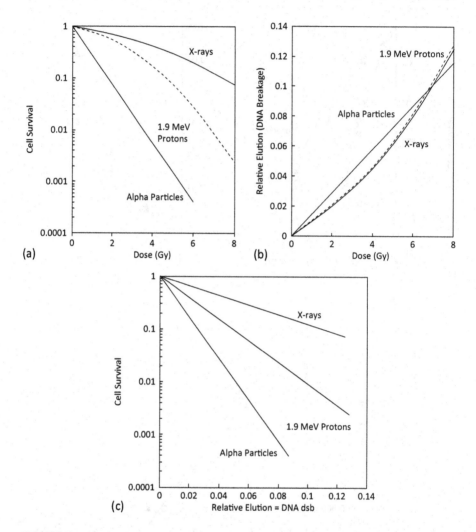

FIGURE 3.8 Cell survival (a) and elution (b) as a function of dose for X-rays, 1.9 MeV protons and alpha particles and the changing linear correlation (c) between DNA double strand breaks and survival for the more densely ionising radiations (redrawn from Prise 1994).

transfer) of 20 keV/μm, which is comparable with the 1.9 MeV protons having an LET of 17 keV/μm. Figure 3.9 presents the correlation between relative elution and cell survival, demonstrating that the slope of the correlation is close to, if not the same as, that found for 250 kVp hard X-rays.

This implies that the alteration of the slope of the correlation between the DNA double strand breaks and the logarithm of cell survival, found for the proton and alpha particles, is not due to the change in LET between the gamma rays and the protons and alpha particles. It might suggest that the non-random induction of DNA double strand breaks in clusters by the protons and the alpha particles does have some role to play in the alteration of the correlation slope.

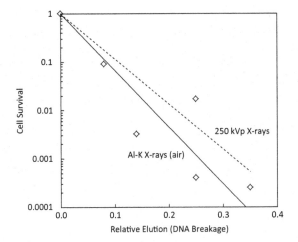

FIGURE 3.9 The correlation between relative elution (DNA double strand breaks) and the logarithm of cell survival after exposure in air to ultrasoft aluminium-K X-rays having an LET of 20 keV/μm. The dashed line is the correlation found for hard X-rays shown in Figure 3.6 (data from Prise et al. 1989).

This means that little importance can be attached to the absolute values of the yields of double strand breaks determined by Prise et al. (1987, 1990; Prise 1994) and others (Weber and Flentje, 1993; Peak et al. 1991; Jenner et al. 1993) after densely ionising radiation. The changing slope of the correlation between DNA double strand breaks and cell inactivation, found after densely ionising radiations by Prise (1994) is, therefore, not surprising and does not invalidate the direct association between double strand breaks and cell inactivation shown in the correlations for sparsely ionising radiation in Figures 3.2 to 3.7.

3.3　THE CORRELATION BETWEEN CHROMOSOMAL ABERRATION YIELD AND CELL SURVIVAL

In Chapter 2, Equation 2.2 related the yield (Y) of chromosomal aberrations to the number of DNA double strand breaks as

$$Y = cN = c\left(\alpha D + \beta D^2\right),$$

which can be rewritten as

$$N = Y/c. \tag{3.3}$$

Combining Equation 3.2 for cell survival with Equation 3.3 to eliminate (N) gives

$$\ln S = -pY/c, \tag{3.4}$$

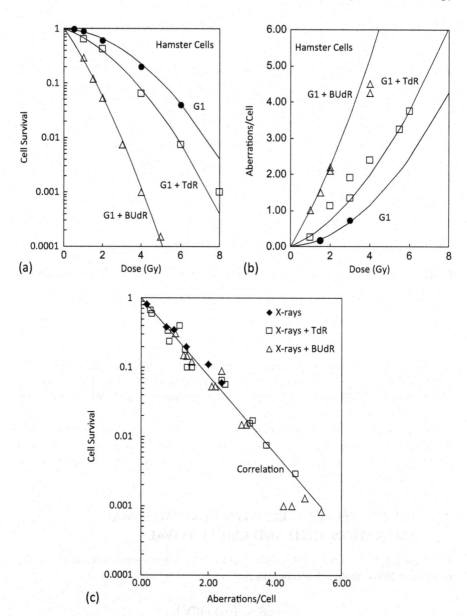

FIGURE 3.10 Dose–effect relationships for cell survival (a) and chromosome aberrations per cell (b) in G1 phase Chinese hamster cells exposed to X-rays with and without a sensitizer. The straight line correlation (c) shows all of the measurements of survival and chromosome aberrations per cell made by Dewey et al. (1970, 1971a,b) in different phases of the cell cycle under the different exposure conditions.

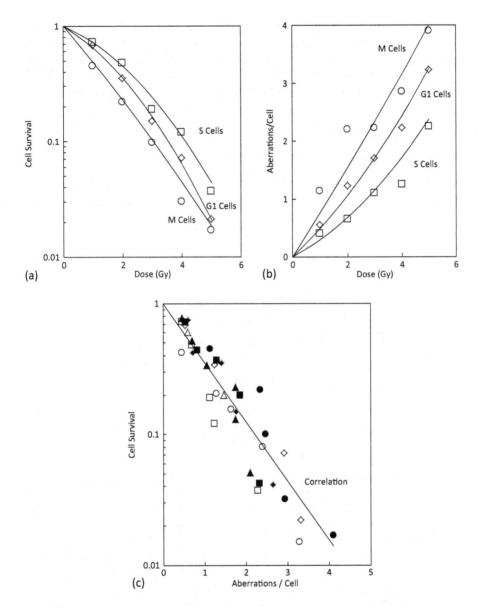

FIGURE 3.11   The upper figures show dose–effect relationships for cell survival (a) and chromosome aberrations per cell (b) for Chinese hamster cells in different phases of the cell cycle synchronised by shake-off. The straight line correlation (c) shows all the measurements of cell survival and chromosome aberrations per cell made by Bhambhani et al. (1973) in different phases of the cell cycle using two different treatments to synchronise the cells. Open symbols used mechanical shake-off, closed symbols used colcemid. Symbols are as identified in the dose–effect figures; triangles are S/G2 cells.

which defines a linear relationship between the logarithm of cell survival and the yield of chromosomal aberrations per cell when the measurements are made on the same population of exposed cells. The linear relationship is independent of the values of the linear (α) and quadratic (β) coefficients which means that the same linear correlation should be found for different exposure conditions in the same cell system.

Dewey et al. (1970, 1971a,b) have made comprehensive measurements of chromosomal aberrations and cell survival in Chinese hamster cells exposed to X-rays in different phases of the cell cycle and under different exposure conditions, including the sensitizer BUdR. Dewey and his colleagues investigated the correlation between the number of chromosome breaks per cell and cell survival and found a good correlation up to three breaks per cell. However, they used the Classical Theory of chromosome aberration formation and counted one break for deletions, two breaks per exchange aberration and four breaks for the complex tri-radial aberrations. In the model shown in Figure 3.1, a DNA double strand break is counted as a chromosome backbone break and deletions and exchange aberrations are counted as one break, with tri-radials counted as two breaks. In Figure 3.10, a representative sample of the dose–effect relationships measured by Dewey et al. (1970, 1971a,b) for cell survival and chromosome aberrations per cell are shown together with the correlation between all their cell survival and chromosome aberration data counted according to the model.

The correlation is good and statistically highly significant up to five aberrations per cell in accordance with Equation 3.4 (Chadwick and Leenhouts 1974, 1981). This correlation coupled with the correlation between DNA double strand breaks and cell survival means that chromosome aberrations must arise from a single chromosome backbone break. (Details of how one chromosome backbone break can lead to exchange aberrations are presented in Chapter 4.) Dewey et al. (1978) have also shown a similar correlation between cell survival and chromosome aberrations per cell after hyperthermia treatment of the cells.

Bhambhani et al. (1973) have also made a comprehensive study of the relationship between chromosome aberrations and cell survival in Chinese hamster ovary cells, synchronised in different phases of the cell cycle using two different methods of synchronisation, either colcemid or 'shake-off'. The correlation between aberrations per cell and cell survival for all their results is shown in Figure 3.11.

The two correlations shown in Figures 3.10 and 3.11 provide a compelling direct link between chromosomal aberrations and cell survival, in accordance with Equation 3.4. The direct link from cell survival to DNA double strand breaks, presented in Figures 3.2 to 3.7, means that a link has been established from DNA double strand breaks to chromosomal aberrations as outlined in the model and illustrated in Figure 3.1.

3.4 THE CORRELATION BETWEEN MUTATION FREQUENCY AND CELL SURVIVAL

The correlation between mutation frequency and cell survival can be derived in a similar way as that for the correlation between chromosome aberrations and cell survival but with a small twist. It has to be realised that very many, if not all, the specific mutations that are measured are associated with a chromosomal aberration.

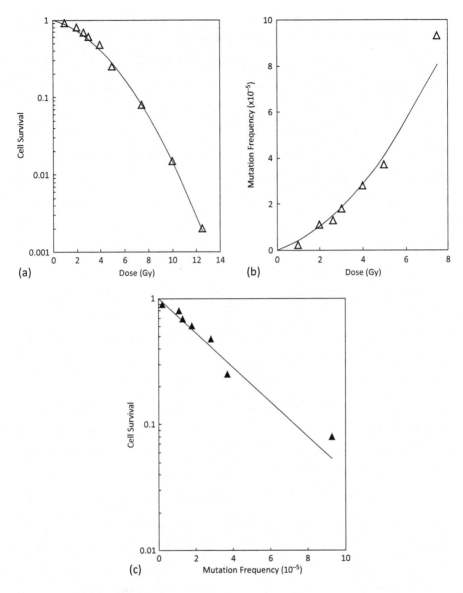

FIGURE 3.12 The dose–effect relationships for cell survival (a) and the induction of 6-thio-guanine-resistant mutations (b) in Chinese hamster cells after gamma irradiation (Thacker et al. 1977) and the straight line correlation (c) between the mutation frequency and survival, in accordance with Equation 3.6.

In Chapter 2, Equation 2.3 described the probability for mutation induction per surviving cell as

$$M = \left(1 - \exp\left(-q\left(\alpha D + \beta D^2\right)\right)\right).$$

It was also explained in Chapter 2 that at frequencies of less than 0.25 mutations per cell this equation could be closely approximated by Equation 2.4, namely,

$$M \approx q\left(\alpha D + \beta D^2\right) = qN.$$

Thus,

$$N = M/q. \tag{3.5}$$

Combining Equation 3.2 for cell survival with Equation 3.5 for mutation frequency per surviving cell to eliminate (N) gives

$$\ln S = -pM/q. \tag{3.6}$$

Equation 3.6 defines a linear relationship between the logarithm of cell survival and the mutation frequency per surviving cell, when the measurements are made on the same population of exposed cells. The linear relationship is independent of the values of the linear (α) and quadratic (β) coefficients which means that the same linear correlation should be found for different exposure conditions in the same cell system.

Figure 3.12 presents data from Thacker et al. (1977) on the induction of 6-thioguanine resistant mutations and survival in Chinese hamster cells after gamma radiation.

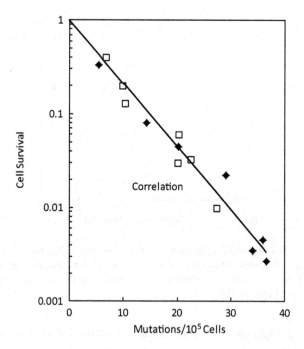

FIGURE 3.13 The straight line correlation between 6-thioguanine-resistant mutations and cell survival in plateau phase hamster cells, after immediate plating (closed diamonds) and delayed plating (open squares) for recovery from potentially lethal damage (Rao and Hopwood 1982), in accordance with Equation 3.6.

Figure 3.13 presents the correlation between mutation frequency and cell survival in stationary phase Chinese hamster ovary cells after gamma radiation before and after a recovery period (Rao and Hopwood 1982) and Figure 3.14 presents a similar correlation in plateau phase mammalian cells after X-rays (Iliakis 1984a,b).

These results are interesting, not only because they are in accordance with Equation 3.6, but also because they show that the increase in survival in the delayed plating period of recovery of the stationary cells is reflected perfectly in the reduction of the mutation

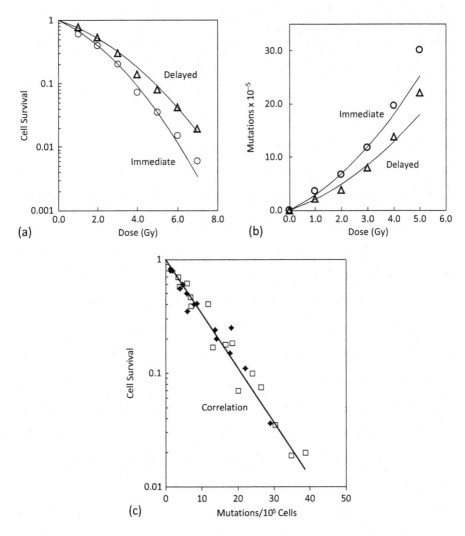

FIGURE 3.14 Dose–effect relationships for cell survival (a) and mutation frequency (b) in plateau-phase mammalian cells on immediate and delayed plating after exposure, and the correlation (c) between 6-thioguanine-resistant mutations and cell survival after immediate plating (closed diamonds) and delayed plating (open squares) for recovery from potentially lethal damage (Iliakis 1984a,b), in accordance with Equation 3.6.

frequency. This suggests that the recovery in survival and reduction of mutations result from the same process, that is, the correct repair of DNA double strand breaks.

The group of Goodhead, Thacker and Cox (Goodhead and Thacker 1977; Cox, Thacker and Goodhead 1977; Goodhead, Thacker and Cox 1979; Goodhead 1980; Cox and Masson 1979; Thacker, Stretch and Stephens 1979; Goodhead, Munson, Thacker and Cox 1980) carried out a far-reaching project examining the survival and induction of 6-thioguanine resistant mutations in V79 Chinese hamster cells and human fibroblast cells exposed to several different radiation types from sparsely ionising cobalt-60 gamma rays to densely ionising helium and boron ions as well as aluminium-K and carbon-K ultrasoft X-rays. The correlation between mutations per surviving cell and the logarithm of survival were investigated using Equation 3.6. The human fibroblast cells are, unfortunately, not very informative about the correlation between mutation frequency and survival because the dose–effect curves for hard X-rays are closely linear and a linear logarithm survival curve plotted against a linear mutation frequency curve always gives a linear correlation relationship. However, as shown in Figure 3.12 the Chinese hamster cells exhibit a curvilinear survival and mutation frequency dose–effect relationship and a correlation between the two endpoints which is in agreement with Equation 3.6.

The results of this project revealed that the linear (α) coefficient of the dose–effect relationships for both cell survival and mutation frequency increased as the more densely ionising radiations were used as anticipated. The dose–effect relationships for the more densely ionising radiations became dominated by the value of the linear (α) coefficient and became almost linear with dose in the Chinese hamster cells as well as the human fibroblasts. All this is in line with the expectations from the model but an interesting and unexplained result found in both cell types, although less obvious in the human fibroblast cells, was a change in the slope of the correlation curves as the more densely ionising radiations were used (see Figure 3.15).

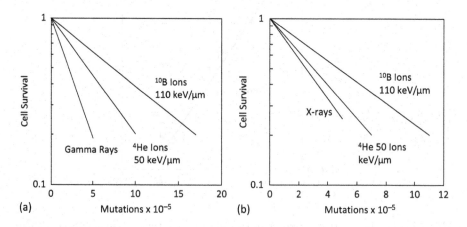

FIGURE 3.15 The correlations between mutation frequency and survival found for different radiations in V79 Chinese hamster cells (a) and human fibroblast cells (b). There are more mutations per survivor as the radiation becomes more densely ionising (data from Goodhead 1980; Goodhead et al. 1979).

The more densely ionising radiations induce more mutations per surviving cell than the sparsely ionising radiations. This is in contradiction with the expectation of the model, which is that the correlation should not change, and raises the question of whether there is an explanation for this contradiction.

As previously mentioned, densely ionising radiation induces clusters of DNA double strand breaks (Rydberg 1996, 2001) which means that the double strand breaks are not induced at random throughout the cell population and this will alter the application of Poisson statistics used for the derivation of cell survival. Whether this has some influence on the correlations remains to be investigated.

However, the group of Goodhead, Thacker and Cox also investigated the induction of mutation and the inactivation of cells using ultrasoft X-rays from aluminium and carbon. These are particularly interesting because the secondary electrons from the aluminium-K X-rays have an energy of 1.5 keV, a track length of 70 nm and an LET (Linear Energy Transfer) of 21 keV/μm. The secondary electrons from the carbon-K X-rays have an energy of 0.28 keV, a track length of 7 nm and an LET of 40 keV/μm. These ultrasoft X-rays and, especially the carbon-K X-rays giving electrons with a track length of 7 nm, close to the dimensions of the DNA double helix at 2 nm, were found to be very efficient at cell inactivation and mutation induction and on a parallel with helium ions having the same LET (see Figure 3.16).

The efficiency of the carbon-K X-rays with an electron track length of only 7 nm defined the size of the target for the cellular endpoints of mutation induction, survival and also chromosomal aberrations as 7 nm or less. This contradicts many other radiation biological models.

However, although these ultrasoft X-rays have LET values comparable with densely ionising radiations, such as the helium ions, the lesions they produce should

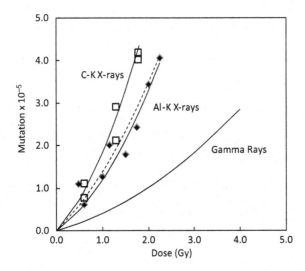

FIGURE 3.16 The efficient induction of mutations in V79 Chinese hamster cells by ultrasoft aluminium-K and carbon-K X-rays compared with the induction by cobalt-60 gamma rays and helium ions with an LET of 20keV/μm (dashed line) which is comparable with the LET of the aluminium-K X-rays (data from Cox et al. 1977).

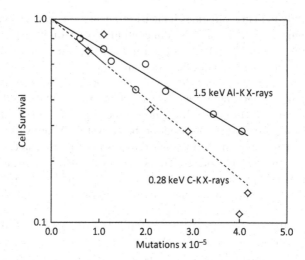

FIGURE 3.17   The correlations between mutations and the logarithm of cell survival for ultrasoft aluminium-K (1.5 keV) and carbon-K X-rays (0.28 keV), according to Equation 3.6. The solid line through the aluminium-K X-rays data is the same as that shown in Figure 3.12 for gamma rays. The dashed line correlation through the carbon-K X-ray data is slightly steeper than that for gamma rays (data from Goodhead et al. 1979 and Cox et al. 1977).

be much more randomly induced in the exposed cell population, so an examination of the correlations induced by these radiations might be informative.

Comparison of Figures 3.15 and 3.17 reveals that the ultrasoft X-rays with LET values similar to the helium ions have the same or almost the same correlations as that found for gamma rays. In fact, the ultrasoft carbon-K X-rays have a slightly steeper correlation, that is, slightly fewer mutations per survivor than the gamma rays, while the helium and boron ions exhibit shallower correlations, that is, considerably more mutations per survivor than the gamma ray correlation. This indicates that the alteration of the slope of the correlation between mutation frequency and the logarithm of cell survival, found for the helium and boron ions, is not due to the change in LET between the gamma rays and the helium and boron ions. It might suggest that the non-random induction of DNA double strand breaks in clusters by the helium and boron ions does have some role to play in the alteration of the correlation slope.

The correlations shown in Figures 3.12 to 3.17 provide a compelling link between mutation frequency per surviving cell and cell survival, in direct accordance with Equation 3.6 and this, together with the correlations between cell survival and DNA double strand breaks, presented in Figures 3.2 to 3.7, means that a link has been established from DNA double strand breaks to mutations, as outlined in the model and illustrated in Figure 3.1.

3.5 AN IMPLIED CORRELATION

There is one more 'implied correlation' that arises in the data on the mutation frequency per irradiated cell (M_I) which is given by Equation 2.5, namely,

$$M_I = M \cdot S = \left(1 - \exp\left(-q\left(\alpha D + \beta D^2\right)\right)\right)\exp\left(-p\left(\alpha D + \beta D^2\right)\right). \qquad (3.7)$$

This is typically the situation for mutations arising as hereditary effects following irradiation of the cells involved in the reproductive process in which an organism is exposed. The mutated reproductive cell has to survive and pass through meiosis to express the mutation.

Figure 2.5 demonstrated that this equation describes a curve which rises to a peak and decreases at higher doses as the effect of cell inactivation starts to dominate. The 'implied correlation' which arises, is that the peak height of the equation is constant and independent of the values of the linear (α) coefficient and the quadratic (β) coefficient. The peak height only depends on the values of (p) and (q). The peak height can be determined from Equation 3.7 by differentiating it with respect to dose (D) and equating the differential to zero. This gives the height in the peak as

$$M_I = \left(1 - p/(p+q)\right)\left(p/(p+q)\right)^{p+q}. \qquad (3.8)$$

This same peak height is extremely well demonstrated in the data of Underbrink et al. (1975) on the induction of pink mutations in the stamen hairs of *Tradescantia* by X-rays and fast neutrons, shown in Figure 3.18, which is drawn with a logarithmic scale on both axes.

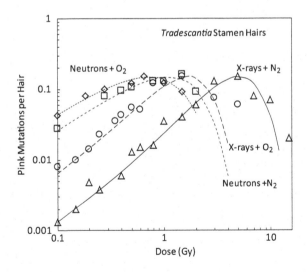

FIGURE 3.18 The induction of pink mutations in stamen hairs of *Tradescantia* by X-rays and fast neutrons in aerobic and anaerobic conditions, illustrating the same peak height 'implied correlation' phenomenon in all four of the dose–effect curves (from Underbrink et al. 1975). Please note that both axes are on a logarithmic scale.

The same peak height 'implied correlation' arises because the same dose kinetics, in terms of the linear (α) and quadratic (β) coefficients, are found in both the induction of mutations and in the cell inactivation parts of Equation 3.7 in each of the four different exposure treatments (Leenhouts and Chadwick 1978a; Chadwick and Leenhouts 1981). In fact, Thorne (2011) has shown that this same peak height phenomenon can be generalised and is valid for a monotonically increasing function of dose irrespective of the actual dose function. This generalisation does not affect the linkage which has been established between mutations, chromosomal aberrations and cell inactivation, and DNA double strand breaks, as shown in Figure 3.1 (Chadwick and Leenhouts 2011b).

Another example of this same peak height is shown in Figure 3.19 for the induction of HGPRT mutations in Chinese hamster cells. The curves have been derived from the data of Thacker et al. (1977) and Cox et al. (1977), shown in Figures 3.12 and 3.17, by multiplying the mutation frequency per surviving cell by the cell survival, in accordance with Equation 3.7.

The importance of the 'implied correlation' will become apparent in Chapter 7 on radiation-induced cancer, where the organism is exposed and the malignantly mutated cell has to survive to express the malignant phenotype. It is also typically the situation that will arise in hereditary effects where an organism is exposed and the reproductive cell has to survive and pass through meiosis to express the hereditary change. This will be discussed in Chapter 9 on radiological protection, but it is worth noting here that the mutation induction per irradiated cell for the specific 6-thioguanine-resistant mutation barely exceeds one mutation in 100,000 cells in

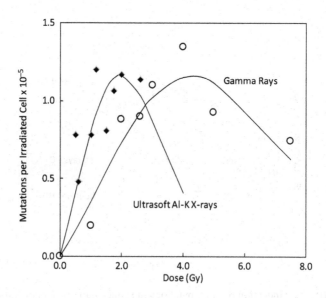

FIGURE 3.19 The induction of 6-thioguanine-resistant mutations in V79 Chinese hamster cells for gamma rays and ultrasoft aluminium-K X-rays, illustrating the same peak height 'implied correlation' (data from Thacker et al. 1977 and Cox et al. 1977).

the peak, and that many of these mutations have been associated with chromosomal aberrations (Thacker et al. 1979; Cox and Masson 1979).

3.6 CONCLUSIONS

Correlations between DNA double strand breaks and the three cellular effects are predicted from a consideration of the dose–effect equations for each cellular effect. Data are presented to illustrate the direct correlation between double strand breaks and cell survival. Data are also presented to illustrate the direct correlations between cell survival and chromosome aberration yield and cell survival and mutation frequency per surviving cell.

The 'same peak height', predicted by the equation for mutation frequency per irradiated cell, is considered to be an implied correlation.

4 DNA Double Strand Breaks and Chromosomal Aberrations

The previous chapters have provided evidence that a single DNA double strand break can lead to a chromosome aberration. A logical explanation for the formation of chromosome aberrations from a single DNA double strand break is presented based on an extension of the recombination repair of a double strand break developed by Resnick. The Resnick recombination repair of a DNA double strand break in one chromosome relies on its undamaged homologous chromosome, or sister chromatid in G2, as a template for correct repair. The extension proposes that only a small region of homology (micro-homology) is needed to achieve the repair. These regions of micro-homology can be anywhere on any of the chromosomes in the nucleus of the cell. It is demonstrated that every chromosome aberration can be derived from a single DNA double strand break when the resolution of the recombination junction fails. It is also shown that the correct resolution of the recombination junction leads to the correct repair of the double strand break. The region of micro-homology between intact DNA and the double strand break DNA means that the recombination repair can take place in all the phases of the cell cycle. Of the two visible breaks seen in chromosome aberrations, one is radiation-induced and the other is a consequence of repair.

4.1 INTRODUCTION

The correlations presented in Chapter 3 gave a direct confirmation of the linear–quadratic dose–effect relationship for the induction of DNA double strand breaks and linked the double strand break to cell survival. The further correlations presented between cell survival and the yield of chromosome aberrations linked chromosome aberrations to DNA double strand breaks and provided strong evidence for the proposal that a single DNA double strand break can be the origin of a chromosomal aberration. The question is: how? The answer lies in a recombination repair process for DNA double strand breaks.

4.2   RECOMBINATION REPAIR OF DNA DOUBLE STRAND BREAKS

In contrast to the repair of DNA single strand breaks, which is achieved because the undamaged single strand provides a template for the accurate replacement of the complementary bases in the correct sequence as the broken strand is restored, a broken DNA double helix does not have a template to copy. Nevertheless, the repair of radiation-induced DNA double strand breaks is a well-established process, although it cannot be assumed to be a perfect repair in each case.

The unineme concept of chromosome structure, which visualises the DNA double helix running from one telomere end of the chromosome to the other telomere end, identifies a DNA double strand break as a chromatid arm break and its repair as the rejoining of chromatid arms. The chromosome carrying the double strand break has apparently no template to copy for the repair. Errors in the rejoining of chromatid arms lead to the formation of chromosomal aberrations which can be seen in the microscope at mitosis.

In 1976, Resnick (1976) outlined a recombination process for the repair of DNA double strand breaks which takes advantage of a sister chromatid or the homologous chromosome as a template and makes use of exchange processes that are known to occur between homologous chromosomes in reproductive cells in meiosis and which shuffle genes from one parental chromosome to the other. Figure 4.1 presents Resnick's recombination repair of a DNA double strand break in a series of steps.

At the double strand break (2 stars, pale grey DNA), exonuclease partially degrades each one of the two DNA strands in the 5′P–3′OH direction to make two

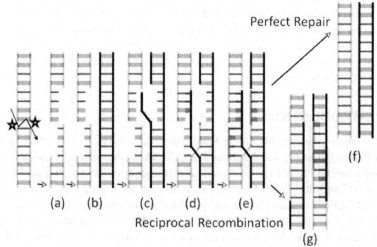

Resnick's Recombination DNA double strand break repair

FIGURE 4.1 A schematic representation of the recombination repair of a DNA double strand break, according to Resnick (1976). The steps are described in detail in the text. There are two possible results of the process: either perfect repair or reciprocal recombination leads to an exchange of DNA helices.

single stranded tails (a). The single stranded tails are recombinogenic with homologous DNA, either from an undamaged sister chromatid in the G2 phase, or with an undamaged homologous chromosome (black DNA) (b). The homologous association at the single stranded tails leads to an endonuclease 'nick' in one strand of the undamaged (black) DNA which allows the strand to pair, by matching complementary bases, with one of the single strand regions at the double strand break (c). This process creates a recombination hetero-duplex, originally proposed by Holliday (1964). Continued unwinding of the undamaged DNA strand extends the hetero-duplex and allows a second hetero-duplex to form between the second single stranded region at the double strand break (pale grey) and the remaining strand of the homologous DNA (black) (d). The single strand gaps (pale grey) can be repaired by complementary base pairing with the two strands of the homologous DNA (black) (e). There are two possible resolutions of the Holliday junction. The DNA unwinds to give the restoration of the original DNA helices and perfect repair (f) (upper drawing) or there is a reciprocal exchange of DNA between the two DNA helices (g) (lower drawing).

It is important to note that, in this second case of reciprocal exchange, the one radiation-induced DNA double strand break leads, apparently, to two visible breaks via the repair process. The second break arises from the cellular repair of the radiation-induced double strand break.

In Figure 4.2, the Resnick recombination repair is extended to the chromosomal situation by adding telomeres and centromeres to the DNA in Figure 4.1. The processes involved remain exactly the same, although steps (a) and (b) in Figure 4.1 have been amalgamated to step (b) in Figure 4.2. Note that a single DNA double strand break leads to a chromosome aberration.

It is clear, in Figure 4.2, that the undamaged chromosome (black) is not the homologous chromosome to the broken chromosome (pale grey). This is because the requirement for homology between the damaged chromosome and the undamaged chromosome does not necessarily require the homology to run the full length of the chromosomes. It only needs to be a small region of homology (micro-homology) on either side of the DNA double strand break. These regions of micro-homology can be anywhere on any of the chromosomes in the nucleus of the cell. The presence of a substantial number of highly repetitive, closely homologous, DNA regions scattered throughout the cell nucleus (Britten and Kohne 1968; Davidson and Britten 1973; Davidson et al. 1973; Thomas et al. 1970) provides the essential link which permits Resnick's recombinant repair of a DNA double strand break to be extended to describe the formation of radiation-induced chromosomal aberrations (Chadwick and Leenhouts 1978, 1981). In fact, in Figure 4.2, the reciprocal exchange of DNA between the damaged chromosome (pale grey) and the undamaged chromosome (black) leads, in step (g), to a well-known, easily identified dicentric aberration.

The requirement that a region of micro-homology is needed between the damaged chromosome and an undamaged chromosome at the radiation-induced double strand break, to facilitate the recombination repair process, means that the repair process can involve other, undamaged parts of the same damaged chromosome or any other undamaged chromosome. It also means that the recombination repair process can take place perfectly well in G1 phase cells, as well as in other phases of the cell cycle.

4.3 REPAIR OF DNA DOUBLE STRAND BREAKS AND THE FORMATION OF CHROMOSOMAL ABERRATIONS

In the following diagrams, a series of chromosomal aberrations is presented as an example of how the chromosomal aberrations arise from this extension of Resnick's repair process. In the G1 phase of the cell, the actual aberration formed depends on where the DNA double strand break is induced on the chromosome and, whether the recombination repair occurs between an undamaged part of the same chromosome (black), or an undamaged part of a different chromosome (black). In all cases of an exchange aberration, micro-homology must occur between the DNA around the double strand break and a region of DNA on the undamaged chromosome. Otherwise, the repair cannot take place (see Figures 4.2 through 4.13).

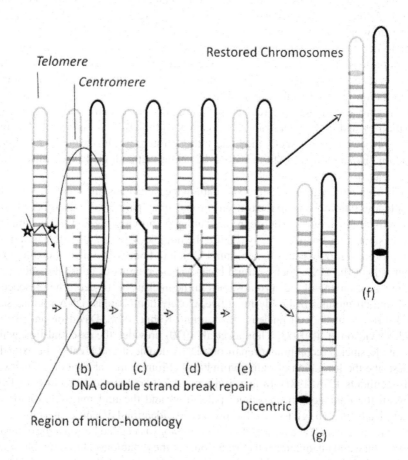

FIGURE 4.2   The Resnick recombination repair of a DNA double strand break extended to the chromosome situation by adding telomeres and centromeres to Figure 4.1. The reciprocal recombination repair can lead to the formation of a dicentric aberration (g) or the restored chromosomes (f). Note the region of micro-homology at the break (not to scale).

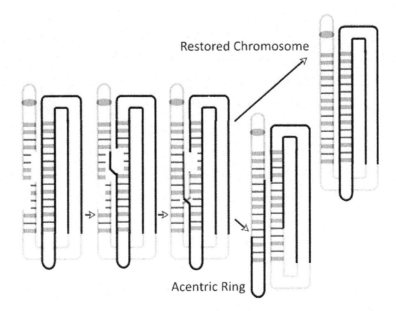

FIGURE 4.3 A schematic diagram of the formation of an acentric ring aberration, following Resnick's reciprocal recombination repair between the double strand break and a region of micro-homology on the same chromosome arm. Note the region of micro-homology at the break (not to scale).

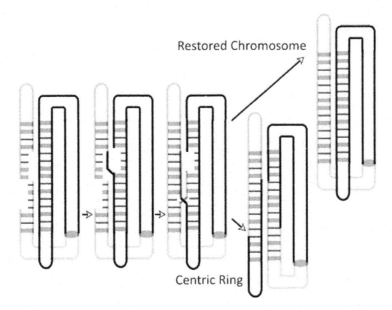

FIGURE 4.4 A schematic diagram of a centric ring aberration formed by recombination repair between the damaged arm and the other arm of the same chromosome. The diagram is very similar to Figure 4.3 but the centromere is positioned differently. Note the region of micro-homology at the break (not to scale).

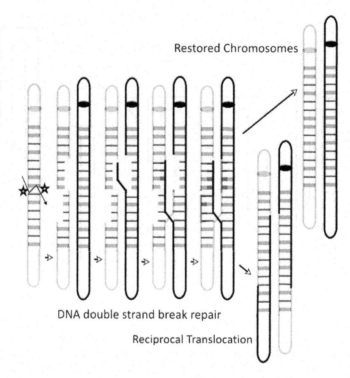

FIGURE 4.5 A schematic diagram of the formation of a reciprocal translocation. The diagram is similar to that for a dicentric aberration in Figure 4.2 but the centromere on the undamaged (black) chromosome is at the top. Note the region of micro-homology at the break (not to scale).

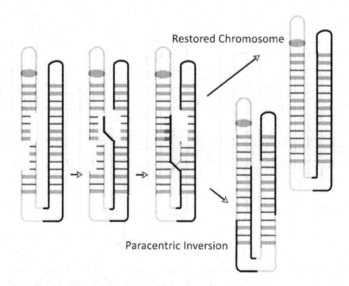

FIGURE 4.6 A schematic diagram of the formation of a paracentric inversion G1 aberration. Note the region of micro-homology at the break (not to scale).

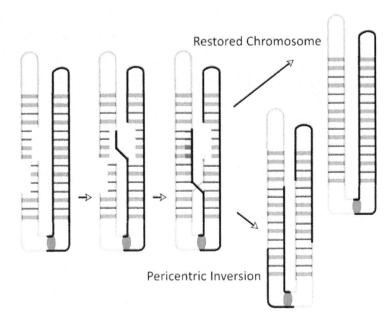

FIGURE 4.7 A schematic diagram of the formation of a pericentric inversion G1 aberration. Note the region of micro-homology at the break (not to scale).

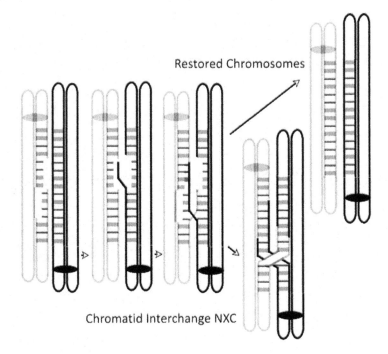

FIGURE 4.8 A schematic diagram of the formation of a chromatid interchange NXC aberration. Note the region of micro-homology at the break (not to scale).

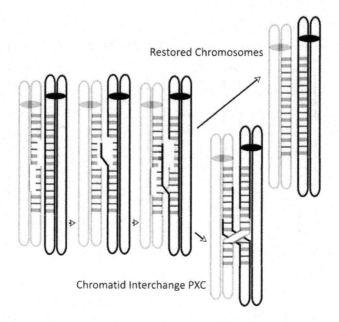

FIGURE 4.9 A schematic diagram of the formation of a chromatid interchange PXC aberration. Note the region of micro-homology at the break (not to scale).

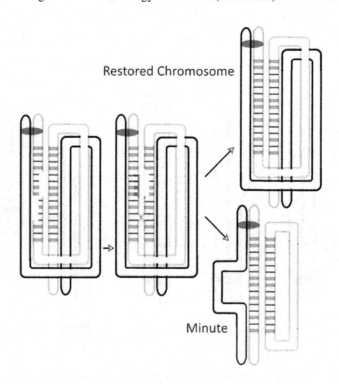

FIGURE 4.10 A schematic diagram of the formation of a minute aberration. Note the region of micro-homology at the break (not to scale).

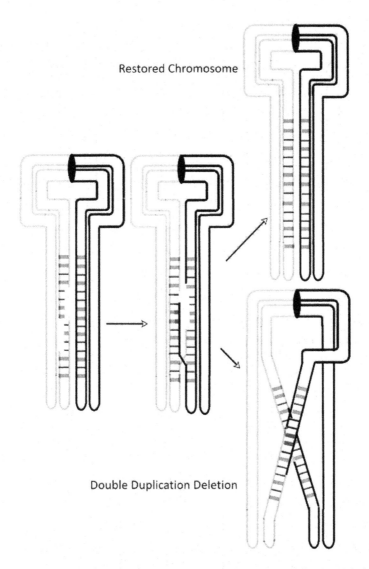

Restored Chromosome

Double Duplication Deletion

FIGURE 4.11 A schematic diagram of the formation of a double duplication deletion aberration. Note the region of micro-homology at the break (not to scale).

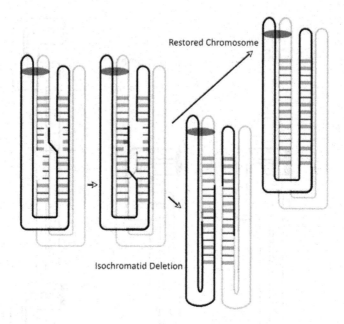

FIGURE 4.12 A schematic diagram of the formation of an isochromatid deletion aberration. Note the region of micro-homology at the break (not to scale).

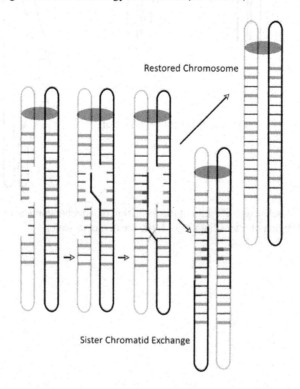

FIGURE 4.13 A schematic diagram of the formation of a sister chromatid exchange aberration. Note the region of micro-homology at the break (not to scale).

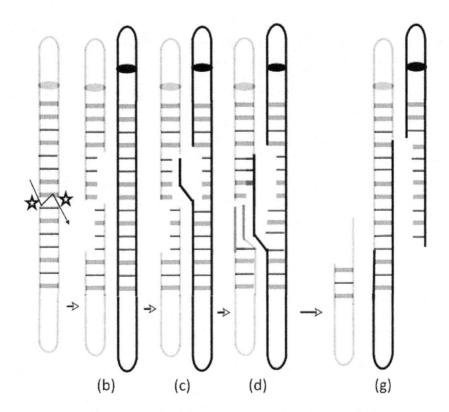

(b)   (c)   (d)   (g)

FIGURE 4.14 A schematic diagram of an incomplete reciprocal recombination repair. The process starts with the formation of the first hetero-duplex at the region of micro-homology (c) but the second hetero-duplex fails to form (d). The first hetero-duplex resolves with the translocation of chromosomal material and two deletions (g).

Figure 4.14 presents schematically a failed reciprocal recombination repair process leading to incompleteness in a reciprocal translocation. The second hetero-duplex fails to form (d) and this leads to the incompleteness, although the first hetero-duplex resolves in a translocation between the two chromosomes. It should not be difficult to visualise the incomplete versions of the other aberrations

An important feature of the recombination repair process of chromosome aberration formation using repetitive DNA micro-homologies, visible in all the diagrams, is that the perfect restoration of the original chromosomes is one of the two possible resolutions of the Holliday type of junction. The perfect restoration of the original damaged chromosome does not depend on the availability of its homologous chromosome for the recombination repair. This is interesting from a radiation biological point of view because, in Chapter 3, it was shown that the delayed plating of stationary cells leads to a reduction in the mutation frequency and a concomitant increase in cell survival (Figure 3.13). It seems reasonable to think of the repair of potentially lethal lesions in stationary cells in terms of Resnick's recombination repair during the delay. Thus, the cell has more time to ensure the correct resolution of the Holliday junction leading to the restoration of the original chromosomes and a consequent reduction in the number of chromosomal aberrations and mutations. This also lends emphasis to the anticipation that most, if not all, specific mutations arise through chromosomal aberrations.

Another important feature of the formation of chromosomal aberrations from the reciprocal recombination repair of a DNA double strand break is that the radiation-induced chromosome break leads, via the repair process, to a second break which can be made visible in the aberration. The second break is not radiation-induced.

If the radiation-induced chromosome break occurs in a unique section of DNA in the G1 phase and there is no region of micro-homologous DNA available for the repair then, either the homologous chromosome of the damaged chromosome might provide the basis for the repair, or a terminal deletion of the damaged chromosome would result. In the G2 phase, the sister chromatid could provide the necessary homology for repair to take place. However, the dose–effect relationship for these terminal deletions and all the other chromosomal aberrations and mutations remains linear–quadratic, in accordance with Equation 2.2, because the induction of the DNA double strand break in the chromosome backbone is linear–quadratic with dose.

In Chapter 3, correlations were presented which provide a direct relationship between the induction of DNA double strand breaks and cell survival and the correlations which link cell survival to the yield of chromosomal aberrations. In this way, a linkage is created from DNA double strand breaks to chromosomal aberrations and that linkage demands a model showing how all the aberrations can arise from a single chromosome backbone break, that is, a single DNA double strand break. Resnick's recombination repair of a DNA double strand break, coupled with the extension of that repair process to the description of chromosomal aberrations via regions of micro-homology on undamaged DNA, completes the circle associating DNA double strand breaks with chromosomal aberrations.

Figures 4.3 through 4.14 demonstrate that all chromosomal aberrations can be produced by the micro-homology-enabled recombination repair (MHERR) process derived from the Resnick recombination repair of a DNA double strand break. The Resnick recombination repair is an organised enzymatic process by which the cell attempts to repair the radiation-induced double strand break damage, taking advantage of the availability throughout the nuclear chromosomes of multiple regions of micro-homology. Even when the undamaged chromosome is not homologous with the damaged chromosome, the process can lead to the perfect repair of the damaged chromosome. However, this depends on the correct resolution of the Holliday junction which cannot be guaranteed to take place and the repair of DNA double strand breaks will never be 100% perfect. Radiation will always leave evidence of its effect.

4.4   STRONG SUPPORTING EVIDENCE

The strongest evidence supporting the concept that a chromosomal aberration could derive from one DNA double strand break comes from the cytological studies using 0.28 keV carbon-K ultrasoft X-rays which deposit their energy in a 7 nanometre long electron track (Goodhead et al. 1979). This electron track is just a little longer than the 2 nanometre DNA double helix is wide but much shorter than the 30 nanometre DNA-histone coil width. Consequently, when the cytological experiments were undertaken, it was widely expected that the dose–effect relationship for chromosomal aberrations would have a very strong quadratic (β) coefficient and a small linear (α) coefficient based on the Classical Breakage–Reunion Theory concepts of aberration formation. In fact, the dose–effect relationships for the chromosomal aberrations, induced by the 0.28 keV ultrasoft X-rays, were very strongly linear with dose (Goodhead et al. 1981; Virsik et al. 1980, 1981; Thacker et al. 1980, 1983, 1986; Simpson and Savage 1996; Griffin et al. 1996; 1998). The 7 nm long electron track is considered unlikely to break two different chromosome arms in its passage through the nuclear DNA. Therefore, it was concluded that the chromosomal aberrations found after the 0.28 keV ultrasoft X-ray exposures must arise from a single DNA double strand break (Goodhead et al. 1981; Virsik et al. 1980; Thacker et al. 1980, 1983, 1986).

Figure 4.15 presents data from Thacker et al. (1983) illustrating the dose–effect relationships for dicentric and ring exchange aberrations after exposure to 0.28 keV ultrasoft X-rays. The linear (α) coefficient for the ultrasoft carbon-K X-rays is three times larger than that for the 250 kVp X-rays.

Another experiment, which offers support for the concept of aberrations arising from a single DNA double strand break, examined aberrations formed between irradiated chromosomes and an unirradiated chromosome arm (Ludwików et al. 2002). This was made possible in synchronised Chinese hamster cells where the Xq chromatid arm was found to replicate after all the other chromosomes had replicated. Iodine-125 labelled iodo-deoxyuridine (IUdR) was added to the cell culture medium until all the chromosomes had replicated except for the late-replicating Xq chromatid

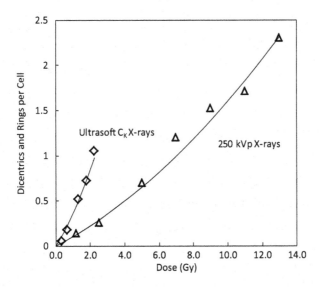

FIGURE 4.15 The number of dicentric and ring aberrations per cell in V79 Chinese hamster cells as a function of dose for cells exposed to ultrasoft 0.28 keV carbon X-rays and 250 kVp X-rays, illustrating the high efficiency of the ultrasoft X-rays at low doses (data from Thacker et al. 1983).

arm. The Xq chromatid arm was the only chromosome arm not containing the IUdR and thus the iodine-125. Iodine-125 emits Auger electrons which have a very short range but which are known to induce DNA double strand breaks at the decay site. The cells were held at the G1/S border to accumulate double strand breaks in the replicated chromosomes. When chromosome aberrations were scored at the first mitosis, exchange aberrations were found between the unexposed, undamaged, late-replicating Xq chromatid arm and the other exposed, damaged chromosomes in the cell nucleus. The authors concluded 'that DNA not damaged by the decay of incorporated ^{125}I can interact with damaged DNA indicating an alternative pathway for the formation of chromosome aberrations' (Ludwików et al. 2002).

Some experimental work has also been done fusing irradiated cells with unirradiated cells and looking for aberrations between the irradiated chromosomes and the unirradiated chromosomes (Cornforth 1990). The study does not appear to show many aberrations formed between irradiated and unirradiated chromosomes but does not rule out the possibility entirely. However, this work appears to have been superseded by the work of Ludwików et al. (2002).

4.5 COMPLEX REARRANGEMENTS

Occasionally, more complex chromosome rearrangements than those illustrated above are found at the first mitosis following irradiation. These complex aberrations appear to have developed from more than the two 'break points' seen in the normal aberrations. In the model for chromosomal aberration formation presented here, these complex rearrangements are considered to arise from two radiation-induced DNA double strand breaks which undergo interaction during the two recombination repair processes. Consequently, a relationship between the yield of the complex rearrangements and the yield of the normal aberrations can be derived.

If the yield of normal chromosome aberrations (Y) is given by Equation 2.2 as

$$Y = c\left(\alpha D + \beta D^2\right)$$

and complex rearrangements arise from two DNA double strand breaks, then the yield (Y_1) of complex rearrangements is

$$Y_1 = c_1\left(\alpha D + \beta D^2\right)^2. \tag{4.1}$$

A correlation between the yield of normal chromosomes and the yield of complex rearrangements, measured in the same experiment, is clearly indicated as

$$Y_1 = c_1 Y^2 / c^2 = kY^2. \tag{4.2}$$

This equation predicts that the yield of complex rearrangements should be proportional to the square of the yield of normal chromosomal aberrations. In Figure 4.16, two examples of the correlation between the yield of complex aberrations and the yield of normal aberrations, predicted by Equation 4.2, are presented.

A similar analysis of some comprehensive data from Revell (1966) was presented in a previous publication (Chadwick and Leenhouts 1981).

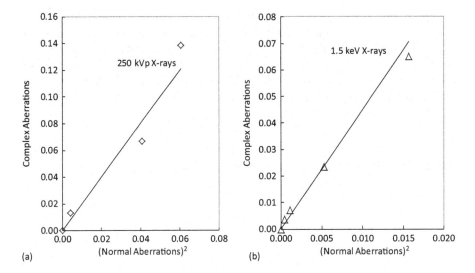

FIGURE 4.16 The correlation between the yield of complex aberrations and the yield of normal aberrations for 250 kVp X-rays (a) (data from Simpson and Savage 1996) and for 1.5 keV X-rays (b) (data from Griffin et al. 1996).

4.6 ADDITIONAL COMMENTS

4.6.1 OTHER ABERRATION FORMATION THEORIES

The linear–quadratic dose–effect relationship for the induction of DNA double strand breaks and the reciprocal recombination repair of those breaks to produce chromosomal aberrations is in contradiction with the Classical Theory of aberration formation (Lea and Catcheside 1942) but some radiation cytologists might see elements of Revell's Exchange Theory of aberration formation (1966, 1974) in the recombination repair process.

Revell compared the exchange process which he observed occurring in aberration formation with the meiotic exchange process. He invoked the vague term 'primary event of damage' and considered that two of these 'primary events' were needed to give the linear–quadratic dose–effect relationships which he found for all chromosomal aberrations. The difference here is that all chromosomal aberrations are considered to arise from a single DNA double strand break, that is, a single chromosome arm break, induced with a linear–quadratic dose–effect relationship, followed by an enzymatically controlled, reciprocal recombination repair, exchange process with an undamaged chromosome.

4.6.2 REPETITIVE DNA

The crucial importance of the presence of a large number of small regions of micro-homology in the form of repetitive DNA (Britten and Kohne 1968; Davidson and Britten 1973; Davidson et al. 1973; Thomas et al. 1970), scattered throughout the chromosomes, should not be underestimated. The extension of Resnick's (1976) recombination repair of a DNA double strand break to describe the creation of chromosome aberrations depends upon its availability. The formation of chromosomal aberrations after radiation exposure is a potential consequence of the cell's attempt to repair the radiation-induced damage. Repetitive DNA should, therefore, not be viewed as 'junk DNA' (Ohno 1972; Makalowski 2003) but as essential for the repair of radiation-induced DNA damage and, possibly, double strand damage in general.

4.6.3 Non-Homologous End Joining and Micro-Homology-Enabled Recombination Repair

There appears to be a widely held opinion that homologous recombination (HR) is inhibited in the G1 phase of the cell cycle (Arnoult et al. 2017). Consequently, the DNA repair process known as 'non-homologous end joining' (NHEJ) is considered to be the important pathway for the repair of DNA double strand breaks in this phase of the cell cycle (Lieber 2010). Although there may be good reasons why homologous recombination (HR) is inhibited in G1, there are no reasons to reject the possibility of the micro-homology-enabled recombination repair process of DNA double strand breaks taking place in G1.

Non-homologous end joining derives from the Classical Theory of chromosome aberration formation (Lea and Catcheside 1942) which was developed to explain the linear–quadratic dose–effect relationship found for aberrations. The aberrations could be seen in the microscope to result from two visible chromosome breaks. The linear–quadratic dose–effect relationship was explained by assuming that each break in a chromosome was linearly proportional with radiation dose and that the combination of the two breaks would be linear–quadratic. The Classical Theory of chromosome aberration formation was developed long before anything was known about the DNA backbone structure of chromosomes and before it was known that a chromosome break would be a DNA double strand break. The concept of non-homologous end joining of DNA double strand breaks between two broken chromosomes was conceived to include DNA double strand break repair in the Classical Theory.

The neutral filter elution measurements of the dose–effect relationship of DNA double strand breaks reviewed in Chapter 3 (Radford 1985, 1986; Prise et al. 1987; and Murray et al. 1989) (see Figures 3.2 to 3.7), which showed a linear–quadratic function of radiation dose, contradict the assumption of the Classical Theory that a chromosome break is proportional with dose. The DNA double strand breaks correlated directly with cell survival and, via the correlation between cell survival and the yield of chromosome aberrations (Dewey et al. 1970, 1971a,b; Bhambhani et al. 1973), indirectly with chromosome aberration formation, implying that a single chromosome backbone break was linear–quadratic with radiation dose. Thus, the model for chromosome aberration formation using micro-homology-enabled recombination repair, presented here, offers a different approach to aberration formation which is compatible with these known experimental facts. Consequently, it is proposed that the concept of non-homologous end joining is redundant and should be replaced by the Resnick (1976) reciprocal repair based micro-homology-enabled recombination repair.

The need for micro-homology at the chromosome aberration break points is both a requirement and a prediction of the model process described here and the detection of short regions of micro-homology at the chromosome break points in chromosome aberrations could provide information on this. In this respect, it is interesting to note that while this book was in preparation, Cornforth et al. (2018) published data showing the presence of micro-homology at the break points of a reciprocal translocation. In fact, the model presented here predicts that regions of micro-homology should be found at the break points of all exchange aberrations.

4.7   TELOMERE TO BREAK REJOINING

In previous publications, the process of joining between a chromosome telomere and a DNA double strand break was postulated as a possible mechanism for chromosomal aberration formation (Leenhouts and Chadwick 1974, 1978a; Chadwick and Leenhouts 1974, 1981; Spanjers et al. 1976). This process is no longer thought to be a major pathway for the production of chromosomal aberrations.

Interestingly, a comprehensive review of telomere-DNA double strand break fusions was published by Bailey and Cornforth (Bailey and Cornforth 2007). The essential conclusion appears to be that fusion between telomeres and DNA double strand breaks only really occurs in a DNA repair deficient cellular background. Bailey and Cornforth (2007) present (their Figure 3) the dose–effect relationship for the induction of DNA double strand breaks by measuring DNA double strand break to telomere fusions in DNA repair deficient mouse p53-/- SCID cells (Bailey et al. 2004). Although Bailey and Cornforth consider the dose–effect relationship to be linear with dose, Figure 4.17 presents their results with a linear–quadratic curve

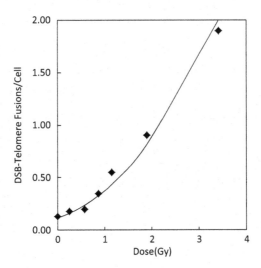

FIGURE 4.17   The dose-response data from Bailey and Cornforth (2007) on DNA double strand break to telomere fusions in DNA repair deficient mouse cells exposed to gamma radiation. The data are fitted with a linear–quadratic curve, in accordance with Equation 1.8.

drawn through the data points. This reanalysis of their data would be consistent with the linear–quadratic dose–effect curves for DNA double strand breaks presented in Chapter 1.

4.8 MUTATIONS

Figure 4.1 presents Resnick's recombination repair of a DNA double strand break and reveals that there are essentially two possible resolutions of the Holliday junction: one that gives a perfect repair of the double helix and one which gives a reciprocal exchange of DNA helices. In Figure 4.2, this process has been generalised for the chromosomal situation and it has been shown that this repair process can lead to either perfectly repaired chromosomes or, following the incorrect resolution of the Holliday junction, all the normally measured chromosomal aberrations. The incorrect resolution of the Holliday junction clearly results in a reorientation of the DNA in different chromosomes which is likely to affect the genetic code and, if the aberrations are not lethal, will cause aberration types of mutations. Both Thacker and Cox (Thacker 1979, 1981; Thacker and Cox 1983; Thacker and Stretch 1983; Thacker et al. 1979; Brown and Thacker 1980; Cox and Mason 1978; Cox et al. 1977) have reported that many of the radiation-induced specific HGPRT mutations measured in V79 Chinese hamster cells and human fibroblast cells are associated with large chromosomal rearrangements and deletions. Kavathas et al. (1980) have reported similar findings in lymphoblastoid cells, and Searle (1974) has reported that specific hereditary mutations in irradiated mice are associated with large genetic alterations. Consequently, mutations associated with chromosome aberrations are probably the most prevalent types.

However, it is worth noting that although the correct resolution of the Holliday junction will normally lead to perfectly repaired chromosomes, it is possible that it can also lead to a small alteration in the genetic coding of one of the chromosomes, as shown in Figure 4.18.

If the short region of micro-homology, between the damaged and undamaged chromosomes used for the repair process, is imperfect and has one or two base pair differences, then the correct resolution of the Holliday junction will lead to slight changes in either the DNA sequence of the radiation damaged chromosome, or the DNA sequence of the undamaged chromosome, in the region of the break.

In either case, this could lead to small changes in the DNA code and this might influence the expression of an adjacent gene or, if the changes occur in an active gene, they might cause a gene mutation. However, the probability for this is likely to be very small indeed, bearing in mind the large amount of 'redundant DNA' in the nucleus and the fact that the level for the induction of HGPRT specific mutations is only around 10^{-5} per surviving cell even though most of these mutations are associated with large chromosomal changes (Thacker and Stretch 1983).

It has to be stressed that the dose–effect relationship for the induction of chromosomal mutations and small coding changes relate back to the dose–effect relationship for DNA double strand breaks which is linear–quadratic with dose. This is

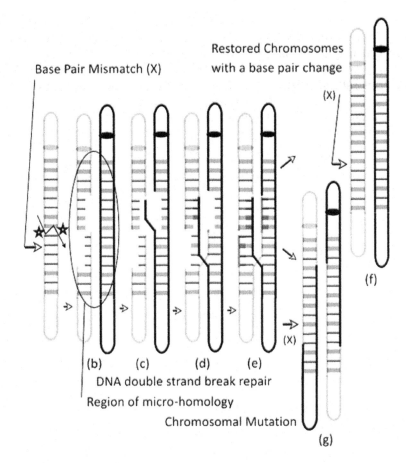

Base Pair Mismatch (X)

Restored Chromosomes
with a base pair change

(X)

(f)

(b)   (c)   (d)   (e)

DNA double strand break repair

Region of micro-homology

Chromosomal Mutation

(X)

(g)

FIGURE 4.18   A schematic diagram of the way the Resnick reciprocal recombination repair of a DNA double strand break can lead to both chromosomal mutations in the form of a reciprocal translocation or, on restoration of the chromosomes, a change in the DNA code of the originally damaged chromosome. A single base pair difference in the region of micro-homology (indicated at (x)) leads to a change in the DNA base pair sequence in the originally damaged chromosome (pale grey).

true for both somatic mutations and hereditary mutations and this linear–quadratic dose–effect relationship has a direct bearing on the risk for health effects arising from somatic mutation, such as cancer, and for the hereditary effects of radiation exposure.

4.9   CONCLUSIONS

The development of the model leads to the prediction that a single DNA double strand break can lead to a chromosome aberration, in contradiction with traditional theories. The formation of chromosome aberrations from one double strand break is described by allowing the Resnick recombination repair of DNA double strand breaks to occur at regions of micro-homology between the damaged chromosome and an undamaged chromosome. All known chromosome aberrations can arise via the recombination repair process. The repair process can lead to the restoration of the damaged chromosome or the formation of an aberration and can take place throughout the cell cycle. Of the two visible breaks seen in chromosome aberrations, one is radiation-induced and the other arises as a consequence of the recombination repair process.

5 The Effect of Dose Rate, Fractionation and Post-Irradiation Repair

The quadratic coefficient of the dose response for double strand breaks arises from the conversion of a 'first' single strand break to a double strand break by a 'secondary' event caused by an independent ionisation track. At low-dose rate, the reduced biological effect that is relevant for radiological risk assessment, results from the cell's ability to repair DNA single strand breaks perfectly during an extended exposure. As the dose rate is reduced, the quadratic coefficient of dose–effect relationships decreases and three regions of exposure times, acute, protracted and chronic, can be defined. The effect of the fractionation of dose is also explained by the repair of the unconverted 'first' single strand breaks, remaining after the first dose fraction, in the time between fractions. An analysis of fractionated dose response relationships is developed. The reduction of effect seen in stationary cells as a result of delayed plating is explained by the repair of DNA double strand breaks during the delay period. This repair is not completely perfect but it is shown that the improved cell survival with delayed plating is accompanied by an equivalent reduction in mutation frequency.

5.1 INTRODUCTION

The explanation of the reduced biological effect of protracted exposure that is relevant for radiological risk assessment, and of the fractionation effect of the dose that is central to radiation therapy regimes in cancer treatment, lies in the cell's ability to repair DNA single strand breaks perfectly. The perfect repair is achieved because the cell has all the enzymatic machinery to replicate its DNA correctly. It can, therefore, reseal a single strand break by complementary base pairing of the undamaged DNA strand across the gap caused by the break.

The explanation of the change in the severity of cellular effects, if the measurements are delayed in stationary cells after exposure, lies in the additional repair of DNA double strand breaks in the delay period, even though this repair is not always perfect.

5.2 THE REPAIR OF DNA SINGLE STRAND BREAKS AND THE DOSE RATE EFFECT

A brief examination of Figure 2.14 and Equation 5.1 for the induction of DNA double strand breaks,

$$N = \left(\alpha D + \beta D^2\right) = \left(2f_0 n\mu k\Omega kD + f_0 f_1 n\mu k(1 - \Omega k)n_1\mu_1 k_1 D^2\right), \tag{5.1}$$

reveals that the induction of DNA double strand breaks in the passage of one ionising particle track (alpha mode, linear coefficient) involves two instantaneous energy depositions, one energy deposition close to each strand of the double helix. There is virtually no time difference between the two energy depositions and, therefore, changes in the duration of exposure will not affect the linear (α) coefficient for the induction of DNA double strand breaks.

The induction of DNA double strand breaks by two independent ionisation particle tracks is not instantaneous, even for very acute exposures, and there will always be a time difference between the induction of the 'first' single strand break and the conversion of that 'first' single strand break to a double strand break, by a 'secondary' event. Consequently, the duration of exposure will affect the quadratic (β) coefficient for the induction of DNA double strand breaks and the process which will influence the quadratic coefficient is the repair of the 'first' DNA single strand break before it is converted to a double strand break. Protraction of radiation exposure means that more time is available for the cell to repair the 'first' single strand breaks perfectly, before they are converted to double strand breaks, by the 'secondary' events. The parameter (f_1) has been defined as the proportion of 'first' single strand breaks which are not repaired before the 'second' break converts them to double strand breaks (see Chapter 1). This parameter, (f_1), which is included in the quadratic (β) coefficient specifically to allow for the repair of DNA single strand breaks, will decrease with protracted exposure. The parameter ($f_1(t)$), accounting for the repair of 'first' single strand breaks, is a function of time, and the average repair time available is one half of the total exposure time.

In an acute exposure, when there is no repair of the 'first' single strand breaks and ($f_1(t)) = 1$, the quadratic component (β) is a maximum and is given by

$$\beta_\infty = n\mu k(1 - \Omega k)n_1\mu_1 k_1. \tag{5.2}$$

As the exposure is protracted, the duration of exposure increases, more 'first' single strand breaks will be repaired before conversion to a double strand break, ($f_1(t)$) decreases until, at very protracted, that is, chronic, exposures, ($f_1(t)$) becomes equal to zero. This means that, as exposure is protracted, the quadratic (β) coefficient will decrease and, at very protracted exposures (β) will become zero.

This is important because it means that, at very protracted exposures, the equation for the induction of DNA double strand breaks is reduced to

$$N = (\alpha D) = \left(2f_0 n\mu k\Omega kD\right). \tag{5.3}$$

The extension of DNA double strand breaks to cause cellular effects such as cell inactivation, chromosomal aberrations and mutations means that, as exposure is protracted, the dose–effect relationships for these three cellular effects will decrease and will become linear functions of dose following very protracted exposures.

This is important for radiological protection because it means that the accumulation of cellular biological effects at very low-dose rates will be directly proportional with radiation dose and there is no threshold dose below which there is no cellular effect.

5.2.1   THREE EXPOSURE TIME REGIONS

Three specific exposure time regions can be defined.

When the exposure time is very short, compared with the repair time for single strand breaks, that is, an acute exposure, then the quadratic (β) coefficient will be maximum and the number of DNA double strand breaks will be

$$N = \left(\alpha D + \beta_\infty D^2\right).\tag{5.4}$$

When the exposure time is very long, compared with the repair time for single strand breaks and all the 'first' single strand breaks are repaired, the quadratic (β) coefficient will be zero and the number of DNA double strand breaks will be

$$N = (\alpha D).\tag{5.5}$$

When the exposure time is comparable with the repair time for single strand breaks, then ($f_1(t)$) will be less than 1 but greater than zero, and the number of double strand breaks will be

$$N = \left(\alpha D + f_1(t)\beta_\infty D^2\right).\tag{5.6}$$

Figure 5.1 illustrates these three exposure time regions, which are defined as 'chronic' when $\beta=0$, 'protracted' when $\beta = f_1(t)\beta_\infty$, and 'acute' when $\beta=\beta_\infty$.

It should be noted that, as the repair of DNA single strand breaks is an enzymatic process, it is temperature dependent (Dugle and Gillespie, 1975) and also dependent

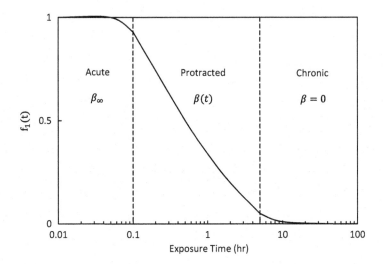

FIGURE 5.1   A diagram of the increasing repair of 'first' single strand breaks when ($f_1(t)$) decreases giving three exposure regions: acute exposure when no repair occurs during exposure; protracted when increasing repair occurs as the exposure time lengthens; and chronic when full repair occurs.

on the metabolic activity and radiation sensitivity of the cells. This means that the actual borders between the three regions are not absolutely defined and might move somewhat to shorter or longer exposure times. Thus, at very short exposure times, the quadratic (β) coefficient will be a maximum, and for very long exposure times, the quadratic (β) coefficient will be zero. In this case, the dose–effect relationship will be defined solely by the linear (α) coefficient.

Figure 5.2 presents one version of the three exposure time regions graphically. The data are derived from an analysis of cell survival measured at different dose rates (Metting et al. 1985; Wells and Bedford 1983) and the figure presents the changing value of the quadratic ($\beta(t)$) coefficient relative to the value of (β_∞) measured for an acute exposure. The survival experiments were made in stationary cells but at constant dose rates and not at constant exposure times, which means that, in order to convert the dose rate to exposure time, a dose of 5 Gy has been assumed to be approximately average in the dose range used. The borders of the three exposure time regions are, consequently, approximate.

However, Thacker and Stretch (1983) have made measurements in plateau-phase Chinese hamster cells to determine the 'acute' region and the 'chronic' region for their experiments on the effect of repair on cell survival and HGPRT mutation induction. They exposed the plateau-phase cells to a dose of 7.78 Gy at a series of different dose rates of cobalt-60 gamma rays, extending from 1.69 Gy/min to 0.078 Gy/hr and measured cell survival. Their results are presented in Figure 5.3 where cell inactivation, defined as (1 – survival), is plotted against the exposure time and reveals the three exposure time regions.

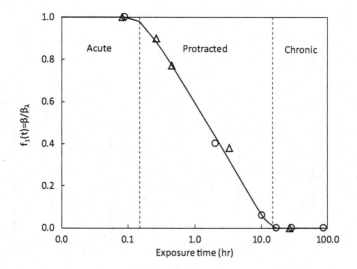

FIGURE 5.2 A presentation of data on cell survival measured at different dose rates in stationary cells (data from Metting et al. 1985 [triangles] and Wells and Bedford 1983 [circles]) illustrating the variation of ($f_1(t)$) as a function of exposure time and the three exposure regions. The dose rates have been converted to exposure time in hours by assuming that a dose of 5 Gy was typical for the exposure range. The exposure time is, consequently, an estimate.

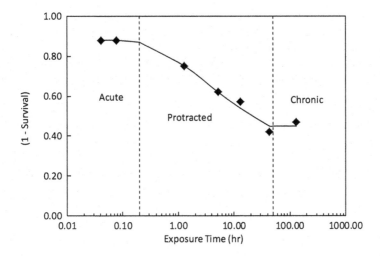

FIGURE 5.3 A plot of cell inactivation (1 – survival) against exposure time for plateau-phase Chinese hamster cells, irradiated with 7.78 Gy of cobalt-60 gamma rays at different dose rates (data from Thacker and Stretch 1983).

It appears from Figures 5.2 and 5.3 that the effect caused by exposures of more than a few hours, at least in stationary cells, will be purely defined by the linear (α) coefficient of the dose–effect curve.

Considerable emphasis has been placed on exposure time rather than dose rate. The reason for this is that the quadratic (β) coefficient is a function of time, as it contains ($f_1(t)$) for the repair of DNA single strand breaks, and is not a function of dose rate. If the repair of single strand breaks is exponential in time, as it appears to be from the data of Dugle and Gillespie (1975), it can be shown that

$$f_1(t) = \left(e^{-\lambda t} - 1 + \lambda t\right) \cdot 2 / (\lambda t)^2 \tag{5.7a}$$

and

$$\beta(t) = \left(e^{-\lambda t} - 1 + \lambda t\right) \cdot 2\beta_\infty / (\lambda t)^2 \tag{5.7b}$$

where (λ) is the repair rate constant for DNA single strand breaks and is related to the 'half-life' for repair (τ) by $\tau = 0.693/\lambda$ (Chadwick and Leenhouts 1981).

Equation 5.7 is independent of dose and dose rate but is dependent on time. The equation is derived for DNA double strand breaks but applies equally to all three cellular effects, cell survival, chromosomal aberrations and mutations.

The function ($f_1(t)$) has this complicated form because the exposure of cells to radiation is a dynamic process and the 'first' single strand breaks are being produced throughout the exposure but, while some are converted to double strand breaks, others are being repaired throughout the exposure.

The time dependence of ($f_1(t)$) means that, in the protracted exposure time region, a dose–effect relationship will have a constant quadratic (β) coefficient $\beta = f_1(t)\beta_\infty$ when each exposure for the dose–effect relationship is given in the same time. This is not what happens when the exposure is given at the same dose rate. Clearly, at a constant dose rate, the exposure time increases as the exposure dose increases.

There are some interesting consequences of this.

If a series of dose–effect relationships is measured using constant exposure times from the acute through the protracted time regions, the different dose–effect relationships should all have the same linear (α) coefficient , and each dose–effect relationship should have a constant value of the quadratic (β) coefficient which should get smaller as the exposure time is increased.

If a series of dose–effect relationships is measured using constant dose rates from acute exposures to protracted exposures, the exposure times of higher doses will increase and in the protracted exposure time region this will allow more repair of the 'first' DNA single strand breaks and the quadratic (β) coefficient will decrease as the dose gets larger. Thus, although the dose–effect relationships will exhibit a sparing effect of reduced dose rate, fitting the different dose–effect relationships, using a linear–quadratic function of dose, will give reduced values of the quadratic (β) coefficient at lower dose rates but the linear (α) coefficients will not remain constant because of 'crosstalk' between the two coefficients.

In fact, with the exponential repair of DNA single strand breaks, it can be shown that, when the dose–effect curve is measured using a constant dose rate in the protracted exposure time region, the dose–effect curve will not be truly linear–quadratic but the curve will have a final constant slope. In the past, dose–effect curves were often analysed with an initial slope and a final slope. The final slope (D_0) of a dose–effect curve made under these conditions can be defined as

$$D_0 = 0.37 / \left(\alpha + 2\beta_\infty D_t / \lambda \right) \qquad (5.8)$$

where (D_t) is the dose rate and (λ) is the repair rate constant for DNA single strand breaks and is related to the 'half-life' for repair (τ) by $\tau = 0.693/\lambda$.

An extensive mathematical assessment of dose rate and repair of single strand breaks can be found in Chapter 7 of *The Molecular Theory of Radiation Biology* (Chadwick and Leenhouts 1981).

In conclusion, the most useful measurements of the sparing effect of extended exposure should be made using constant exposure times for the full dose–effect curve, even though this is not, necessarily, the easiest.

Clearly, if the dose–effect curve does not have a significant quadratic (β) coefficient, such as after exposure to densely ionising radiation when the induction of double strand breaks in the passage of one radiation track is very efficient, there will not be a dose rate effect.

5.3   EXPERIMENTAL MEASUREMENTS OF THE DOSE RATE EFFECT

The experimental investigation of the sparing effect of dose rate is not easy to do because the time of exposure can be long. With a half-life for the repair of DNA single strand breaks in the order of 0.5 hr, an exposure time to ensure the, more or less, total repair of 'first' single strand breaks will be of the order of 30 hours. A good analysis needs a cell system which maintains a constant radiation sensitivity over this period but does not divide (stationary cells), is metabolically active for repair to take place, and has a constant plating efficiency. Ideally, plating of the cells should be delayed for a few hours after exposure to allow for the repair of DNA double strand breaks which might take place during and after the exposure. In addition, care needs to be taken to ensure that the radiation field used for all the different exposures from acute to chronic remains the same.

There are several sets of data on cell survival, chromosomal aberrations and mutations which have examined the effect of dose rate changes on the severity of effects, although most studies have, unfortunately, used constant dose rates rather than constant exposure times. Examples of these studies are presented in the following figures.

5.3.1 CELL SURVIVAL

A detailed study of the dose rate effect on cell survival has been made by Wells and Bedford (1983) who went to some lengths to ensure that the stationary cells used (C3H 10T1/2) satisfied the requirements for a good analysis but, unfortunately, they used constant dose rates. Figure 5.4 reproduces the different survival curves that they measured using dose rates from 55.8 Gy/hr down to 0.06 Gy/hr.

Wells and Bedford found the expected sparing effect of protracted exposure from the acute survival measured at 55.8 Gy/hr down to 0.29 Gy/hr when the survival curve became purely linear with dose, and a further reduction of dose rate down to 0.06 Gy/hr had no effect on cell survival.

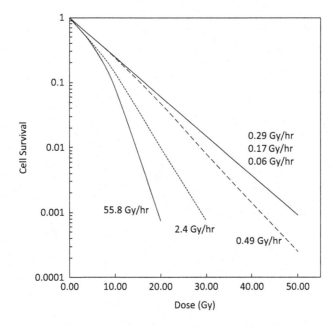

FIGURE 5.4 Cell survival measured in stationary C3H 10T1/2 cells over a wide range of dose rates from 55.8 Gy/hr to 0.06 Gy/hr (curves redrawn from the data of Wells and Bedford 1983) illustrating the sparing effect of reduced dose rate exposure.

It is interesting to note that the fitting of a linear–quadratic dose–effect relationship to the acute survival curve made by Wells and Bedford (1983) defined the value of the linear (α) coefficient equal to 0.15 Gy^{-1} and the average value of the linear slope of the chronic survival curve as D_0 equal to 7.32 Gy which converts to a linear (α) coefficient equal to 0.1366 Gy^{-1}, the same, within the errors, as that for the acute survival curve, exactly as would be expected.

A quick glance at the figure also shows that the two dose–effect relationships measured in the protracted dose rate region show final constant slopes with dose, compatible with the predictions made in Equation 5.8.

In fact, a comprehensive analysis of all the data published by Wells and Bedford (1983) on cell survival in stationary cells exposed to different dose rates can be made, taking advantage of Equation 5.7, coupled with the equation for cell survival:

$$S = \exp\left(-p\left(\alpha D + f_1(t)\beta_\infty D^2\right)\right). \tag{5.9}$$

The analysis, in which only the quadratic ($f_1(t)\beta_\infty$) coefficient is allowed to vary with exposure time, is shown in Figure 5.5 and gives a value of the half-life of repair of single strand breaks of 104 minutes, under the conditions in the stationary cells.

Another detailed study of the effect of reducing dose rate on cell survival was made by Metting et al. (1985) using stationary phase cells. Metting et al. also

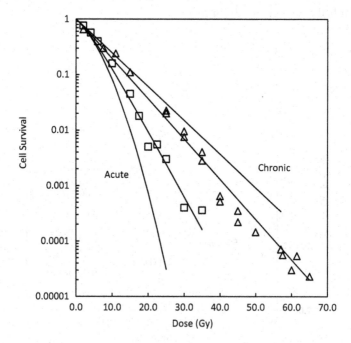

FIGURE 5.5 The analysis of the data of Wells and Bedford (1983) on cell survival following exposures at decreasing dose rates, taking advantage of Equation 5.9 coupled with Equation 5.7. The data shown are for exposure at 2.4 Gy/hr (open squares) and at 0.49 Gy/hr (open triangles).

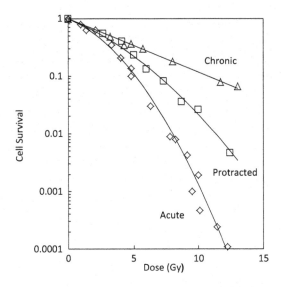

FIGURE 5.6   The dose–effect relationships for cell survival in stationary CHO cells following an acute, protracted and chronic exposure, illustrating the sparing effect of reduced dose rate (from Metting et al. 1985).

investigated the role of delayed plating and fractionation on cell survival. Figure 5.6 presents their dose rate data for cell survival after an acute exposure at 60 Gy/hr, a protracted exposure at 1.5 Gy/hr and a chronic exposure at 0.186 Gy/hr.

5.3.2   CHROMOSOME ABERRATIONS

Edwards and Lloyd (1980) have also discussed the need for dose–effect curves to be measured using constant exposure times. However, their arguments are based on the Classical Theory of chromosomal aberration induction and are only applied to chromosomal aberration yields. Lloyd et al. (1984), subsequently, measured a series of dose–effect relationships for chromosome aberrations in non-cycling human lymphocytes, using constant irradiation times of 1, 3, 6 and 12 hours, to cobalt-60 gamma rays. An acute exposure to cobalt-60 gamma rays made some years prior to the constant time of exposure experiment was also used in the analysis. Precautions were taken to ensure that the unstimulated lymphocytes were metabolically active, maintained at a constant temperature during exposure and cultured using a standard procedure after exposure. Care was also taken to ensure that the radiation field remained constant during the experiment. Figure 5.7 presents the results for dicentric aberrations and for acentric aberrations using the parameters derived by Lloyd et al. (1984).

The figure illustrates the sparing effect of increasing the exposure time. The data for each dose–effect relationship could be closely fitted with a linear–quadratic equation and the linear (α) coefficient of all the curves was closely similar. The chronic linear dose–effect relationship in the figure is derived from the (α) coefficient of the other curves although there were no data.

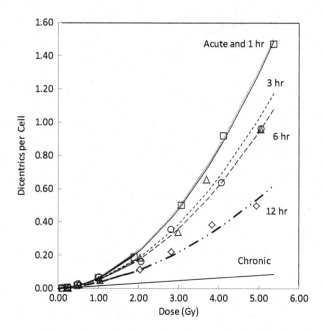

FIGURE 5.7 Dose–effect relationships for dicentric chromosome aberrations measured in human lymphocytes after exposure to cobalt-60 gamma rays with exposure times held constant for each dose–effect relationship (Lloyd et al. 1984). The time of exposure is indicated against the linear–quadratic curves drawn from the results of Lloyd et al. (1984). The open squares are the data for a 1-hour exposure, open triangles for 3 hours, open circles for 6 hours and open diamonds for 12 hours.

5.3.3 Mutation Frequency

The measurement of HPRT mutations in cultured human lymphocytes has been developed as a biomarker of radiation exposure and Vivek Kumar et al. (2006) have studied the effect of dose rate changes on the response of this system.

The radiation sensitivity of the non-cycling human lymphocytes is not expected to change as the exposure duration increases. Although the experiment is far from ideal because the cells were exposed to cobalt-60 gamma rays at room temperature (in India) at the high-dose rate and exposed to caesium gamma rays at 37°C at the low-dose rate, the results do show the sparing effect of increasing the time of exposure both for survival and for mutation induction. Figure 5.8 presents the results for the HPRT mutation induction.

Thacker and Stretch (1983) have also demonstrated the sparing effect of reduced dose rate of cobalt-60 gamma rays on the survival and the induction of HGPRT mutations in plateau-phase V79 Chinese hamster cells. Their results are shown in Figure 5.9 and reveal that at the very low-dose rate used, both the survival and the mutation induction exhibit an almost negligible quadratic (β) coefficient, in contrast to the strong quadratic (β) coefficient found after the acute exposure.

Unfortunately, after the acute exposures, cells were cultivated for measurement immediately. whereas the very low-dose rate exposures (0.2 Gy/hr) meant that these

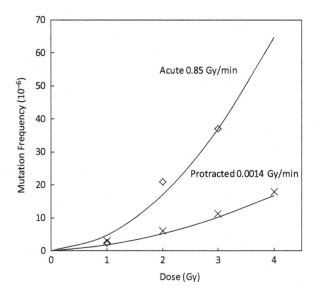

FIGURE 5.8 Dose–effect relationships for HPRT mutations in human lymphocytes after exposure to cobalt-60 gamma rays at 0.85 Gy/min (acute) and to caesium-137 gamma rays at 0.0014 Gy/min (protracted) (Vivek Kumar et al. 2006), illustrating the sparing effect of increased exposure time.

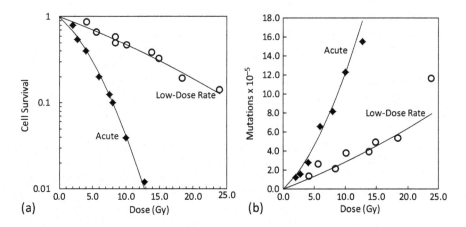

FIGURE 5.9 Cell survival (a) and mutation induction (b) in plateau Chinese hamster cells after acute and chronic exposures to cobalt-60 gamma rays, illustrating the sparing effect of low-dose rate exposures (data redrawn, using mean values of mutation induction from Thacker and Stretch 1983).

cells were exposed for between 20 hours to 5 days before cultivation, during which time post-irradiation repair could take place and this accounts for the fact that the linear (α) coefficient does not remain constant between the acute curves and the chronic curves ($\alpha = 1.59 \times 10^{-1}$ Gy^{-1} compared with $\alpha = 0.66 \times 10^{-1}$ Gy^{-1}).

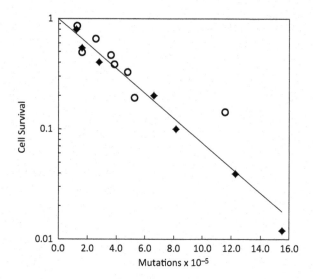

FIGURE 5.10   The correlation between the mutation induction and the logarithm of cell survival following acute exposures (closed diamonds) and chronic exposures (open circles) to cobalt-60 gamma rays (data redrawn, using mean values of mutation induction, from Thacker and Stretch 1983).

Thacker and Stretch (1983) also measured survival after 5 hours of post-irradiation holding before cell cultivation and the linear (α) coefficient of 0.75×10^{-1} Gy^{-1} from the delayed plating survival curve is much closer to that measured at chronic exposure rates. Even so, it is interesting to note that, as shown by Thacker and Stretch (1983), the linear correlation, predicted to exist between the mutation induction and the logarithm of survival in Equation 3.6 in Chapter 3, remains the same for the acute data and the chronic data, indicating an original common molecular mechanism for the induction of both endpoints. This correlation is shown in Figure 5.10.

Another measurement of the effect of protracted exposure on the induction of mutations investigated the frequency of pink mutations in the stamen hairs of *Tradescantia* flowers (Leenhouts et al. 1986). Figure 5.11 presents the data following an acute exposure, a 6-hour exposure and a 12-hour exposure to X-rays. The acute exposure data are fitted to a linear–quadratic dose function and both the 6-hour and 12-hour exposure data lie on the chronic linear dose–effect relationship. The linear (α) coefficient is 6.3×10^{-2} Gy^{-1}, and is the same in both dose–effect curves, as expected. The figure reveals the sparing effect of protraction on the induction of the mutations.

Figures 5.4 to 5.11 demonstrate the sparing effect of either reduced dose rate or extended exposure time on all three cellular effects, cell inactivation, chromosomal aberrations and mutations. The explanation for this sparing effect is based on the perfect repair of 'first' DNA single strand breaks during exposure before they are converted to DNA double strand breaks. It is important to realise that this repair process is, indeed, perfect and there is no influence of the repaired single strand breaks on the cellular effects or on the shape of the dose–effect curves.

The data in the figures suggest that, in some cases, exposures of many hours are required to be in the chronic exposure time region when the dose–effect relationship

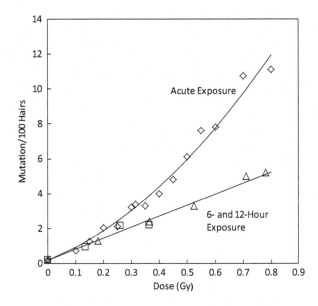

FIGURE 5.11 The dose–effect relationships for the induction of pink mutations in the stamen hairs of *Tradescantia* after acute and protracted (6-hr and 12-hr) exposures to X-rays illustrating the sparing effect of protracted exposure (data from Leenhouts et al. 1986).

is linear and defined by the linear (α) coefficient. The data also show that the linear (α) coefficient of the chronic exposures is the same as the linear (α) coefficient of the comparable acute dose–effect relationship. In the case of constant exposure times, the linear (α) coefficient of the linear–quadratic dose–effect relationships, measured in the protracted exposure time region, is also the same as the linear (α) coefficient found for the comparable acute dose–effect relationship. This is important for radiological protection which is mainly concerned with chronic radiation risk because useful information can be obtained from careful cellular experiments in the acute and protracted exposure regions.

5.4 THE EFFECT OF DOSE FRACTIONATION

Dose fractionation regimes are widely used in radiotherapy treatments of cancer. The linear–quadratic dose–effect relationship offers some interesting analyses of fractionated data which are developed in this section.

The fractionation of the dose, with a suitable time interval between fractions, restores the shoulder of the cell survival curve, as can be seen in Figure 5.12. The dashed straight line in the figure passes through the survival value for one, two and three 4 Gy fractions with full recovery of the shoulder between fractions. This means that these values are related because the survival after two fractions (S_2) is the square of survival (S_1) after one fraction, that is, ($S_2 = S_1^2$) and ($S_3 = S_1^3$). Without full recovery of the shoulder between fractions, the dashed line would not be linear.

The shoulder is recovered because the usual 24-hour interval between fractions permits the repair of the 'first' DNA single strand breaks, induced by the priming dose, which have not been converted into DNA double strand breaks at the end of the

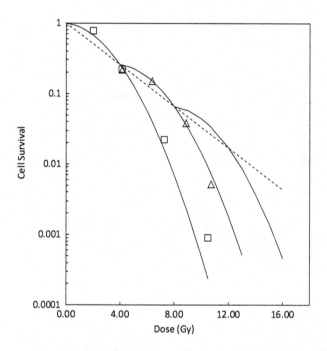

FIGURE 5.12 An example of dose fractionation with a 24-hour interval between fractions (data from Metting et al. 1985). The open squares give the acute dose–effect relationship, the open triangles represent a priming dose of 4 Gy followed by a 24-hour interval and then a further series of acute doses. A second fraction curve at 8 Gy is drawn as an illustration.

first fraction. This means that there is no interaction between the priming dose and the subsequent fraction doses and this can be understood mathematically as follows:

If the total dose is given in two fractions, D_1 and D_2, with no time for repair, then the number of double strand breaks (N) is

$$N = \alpha\left(D_1 + D_2\right) + \beta_\infty\left(D_1 + D_2\right)^2. \tag{5.10}$$

When an interval between fractions permits the full repair of DNA single strand breaks, then the number of double strand breaks (N_F) is

$$N_F = \alpha\left(D_1 + D_2\right) + \beta_\infty\left(D_1^2 + D_2^2\right). \tag{5.11}$$

The difference is the interaction term and is

$$N - N_F = 2\beta_\infty D_1 D_2. \tag{5.12}$$

Interestingly, if the two fraction doses (D_F) are equal, so that ($D_1 = D_2$), then

$$N_F = 2\left(\alpha D_F + \beta_\infty D_F^2\right). \tag{5.13}$$

In general, if a total dose (D) is given in (υ) equal fractions (D_F), with full repair of DNA single strand breaks between fractions, then

$$N_F = \upsilon\left(\alpha D_F + \beta_\infty D_F^2\right) \tag{5.14}$$

and cell survival (S_F) is

$$S_F = \exp\left(-\upsilon p\left(\alpha D_F + \beta_\infty D_F^2\right)\right). \tag{5.15}$$

Since the total dose ($D = \upsilon D_F$), Equation 5.15 can be simplified to

$$-\ln S_F / D = p\left(\alpha + \beta_\infty D_F\right). \tag{5.16}$$

Equation 5.16 predicts that a plot of $\left(-\ln S_F / D\right)$ versus (D_F) will give a straight line of slope ($p\beta_\infty$) with an intercept at ($p\alpha$), for all fraction data satisfying the restrictions imposed by the analysis.

The theoretical analysis of fractionation experiments, which can be developed from the linear–quadratic dose–effect relationship, imposes several restrictions that the experimental protocol must satisfy, if the analysis is to be tested properly.

- The radiation sensitivity of the cells should not change throughout the whole period of the fractionated regime. This can be achieved by using stationary cells.
- The dose rate of the radiation must be sufficiently acute so that all irradiations are short enough to avoid any significant repair of DNA single strand breaks during irradiation.
- All confounding influence of post-irradiation repair of potentially lethal damage, namely, DNA double strand breaks, must be avoided.
- The time between fractions must be long enough to allow the total repair of DNA single strand breaks.

Equation 5.16 is general for all fractionation regimes when the total dose is given in equal fractions and full repair of DNA single strand breaks occurs between fractions. It also introduces an alternative way to present an acute survival curve because it is valid for an acute irradiation when the dose is given in one fraction, that is, when ($D = D_F$) because

$$S = \exp\left(-p\left(\alpha D + \beta_\infty D^2\right)\right) \tag{5.17}$$

and

$$-\ln S/D = p\left(\alpha + \beta_\infty D\right). \tag{5.18}$$

Figure 5.13 presents an acute cell survival curve in both the normal presentation already encountered, and in the form of Equation 5.18.

A comparison of Equations 5.16 and 5.18 reveals that they are directly equivalent. This predicts that all the data from a fractionated series of survival experiments,

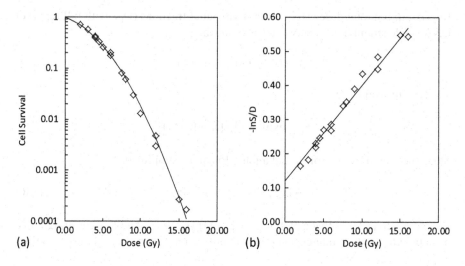

(a) (b)

FIGURE 5.13 Two different presentations of an acute survival curve for stationary CHO cells exposed to gamma rays. The first (a) is drawn according to Equation 5.17, and the second (b) is drawn according to Equation 5.18. The linear (α) coefficient in both cases is 0.12 Gy^{-1} and the quadratic (β) coefficient in both cases is 0.028 Gy^{-2} (data from Chadwick et al. 1998).

including the acute, single-dose data, should lie on the same straight line graph derived from Equation 5.16.

Figure 5.14 presents the right-hand graph of Figure 5.13 but with the data from fractionated exposures inserted in place of the acute survival data.

The data of Leenhouts and Sijsma are all derived from experiments that satisfy the criteria imposed by the analysis. Leenhouts and Sijsma also investigated the

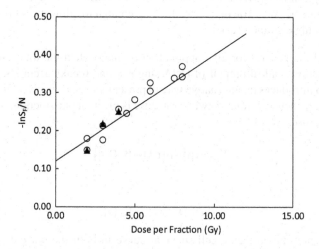

FIGURE 5.14 The linear curve according to Equations 5.16 or 5.18 from Figure 5.13 with data from fractionated exposures of stationary CHO cells with 24 hours between fractions. The open circles are for two fractions covering a total dose range up to 16 Gy and the triangles are for three fractions covering a total dose up to 12 Gy (data from Chadwick et al. 1998).

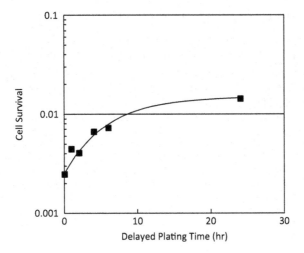

FIGURE 5.15 Survival of stationary CHO cells after an acute dose of 10 Gy with plating of the cells at different times after exposure up to 24 hours (data from Chadwick et al. 1998).

time dependence of delayed plating and the time dependence between fractions on survival to avoid any confounding influence on their fraction experiments (see Figures 5.15 and 5.16).

It should be obvious that, if the dose–effect curve does not have a significant quadratic ($p\beta$) coefficient, such as after exposure to densely ionising radiation when the induction of double strand breaks in the passage of one radiation track is very efficient, there will not be a dose fractionation effect.

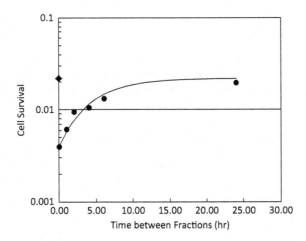

FIGURE 5.16 Survival of stationary CHO cells exposed to two fractions of 6 Gy with different times between fractions. The filled squares are data for the two fractions. The filled diamond is the survival expected after two fractionated doses of 6 Gy with full repair, derived from the square of survival after one 6 Gy dose (data from Chadwick et al. 1998).

5.4.1   THE INFLUENCE OF POST-IRRADIATION REPAIR

In order to avoid any influence of post-irradiation repair in stationary cells affecting the experimental results and, bearing in mind that some of the fractionated experiments can take many hours, cell plating should be delayed after the final exposure for long enough to ensure that no further repair of DNA double strand breaks will occur. The parameter which accounts for the potential repair of DNA double strand breaks after an acute exposure is (f_0).

The time course of the post-irradiation repair can be determined by measuring survival after a fixed, acute exposure of stationary cells, with cell plating made at a series of times after the exposure. Figure 5.15 presents data for stationary CHO cells exposed to 10 Gy of acute gamma rays with plating at different times after exposure.

Figure 5.15 can be analysed using the equation

$$\ln S_D = f_0(t) \ln S \qquad (5.19)$$

where

$$f_0(t) = \exp(-\lambda_d t) + f_0 \qquad (5.20)$$

and (S) is survival after immediate plating, (S_D) is survival after delayed plating, (t) is the delay time, (λ_d) is the repair coefficient of repairable DNA double strand breaks, (f_0) is the fraction of unrepaired or mis-repaired DNA double strand breaks remaining at long delay times. Figure 5.15 gives a repair half-life for DNA double strand breaks in the stationary CHO cells of 4.0 hours. Delayed plating of 24 hours or more will satisfy the restrictions of the fractionated experiments.

5.4.2   THE REPAIR OF DNA SINGLE STRAND BREAKS BETWEEN FRACTIONS

The repair coefficient of DNA single strand breaks between two fractionated exposures can be determined by exposing stationary cells to two acute equal fraction doses with different times between the fractions. Plating of the cells must be delayed by some 24 hours after the second fraction dose to avoid confounding by the post-irradiation repair process. Figure 5.16 presents data for stationary CHO cells exposed to two fractions of 6 Gy with different times between exposures.

Figure 5.16 can be analysed to determine the repair coefficient for DNA single strand breaks using the equation

$$\ln S_t = \ln S_F + (\ln S_0 - \ln S_F)\exp(-\lambda_s t) \qquad (5.21)$$

where (S_0) is survival after two fractions with no time between fractions, (S_F) is the survival expected after two fraction doses of 6 Gy with full repair between fractions, (S_t) is survival after two fraction doses of 6 Gy with time (t) between fractions, and (λ_s) is the repair coefficient for DNA single strand breaks.

Figure 5.16 gives a half-life for the repair of DNA single strand breaks in the stationary CHO cells as 2.9 hours. This is longer than the 40 minutes measured by Dugle and Gillespie (1975), probably because the stationary CHO cells are less metabolically active than those used by Dugle and Gillespie. In the system used by Leenhouts and Sijsma, fractions needed to be separated by more than 12 hours to avoid any interaction between the two doses. Consequently, the two and three fraction data, shown in Figure 5.14, were all made with 24 hours between fractions and a 24-hour post-irradiation period following the last fraction (data is drawn from measurements and analyses performed by Leenhouts and Sijsma [Chadwick et al. 1998]).

5.4.3 FURTHER ANALYTICAL POSSIBILITIES

The use of equal-dose fractions permits two equations to be derived. One equation relates the quadratic (pβ) coefficient to the square of the fractionation dose whereas the other relates the linear (pα) coefficient to the fractionation dose.

The survival of cells after irradiation was defined in Equation 2.1 as

$$S = \exp\left(-p\left(\alpha D + \beta_\infty D^2\right)\right), \tag{5.22}$$

which can also be written in terms of (υ) equal fractions with no repair between fractions as

$$S = \exp\left(-p\left(\upsilon\alpha D_F + \upsilon^2\beta_\infty D_F^2\right)\right). \tag{5.23}$$

If full repair of the DNA single strand breaks occurs between the (υ) fractions, the survival is given by Equation 5.15

$$S_F = \exp\left(-p\left(\upsilon\alpha D_F + \upsilon\beta_\infty D_F^2\right)\right). \tag{5.15}$$

Thus,

$$S_F / S = \exp\left(p\upsilon(\upsilon - 1)\beta_\infty D_F^2\right), \tag{5.24}$$

which gives

$$(1/\upsilon(\upsilon - 1)) \cdot \ln\left(S_F/S\right) = p\beta_\infty D_F^2. \tag{5.25}$$

Equation 5.25 predicts that a plot of $(1/\upsilon(\upsilon - 1)) \cdot \ln$ (S_F/S) against (D_F^2) will give a straight line of slope (pβ_∞) for all fractionation data which satisfy the restrictions imposed by the analysis (see Figure 5.17).

The linear (pα) coefficient can be derived in a similar way from Equations 5.23 and 5.15, if Equation 5.15 is raised to the power of (υ), the number of fractions.

Then

$$\left(S_F\right)^\upsilon = \exp\left(-p\left(\upsilon^2\alpha D_F + \upsilon^2\beta_\infty D_F^2\right)\right) \tag{5.26}$$

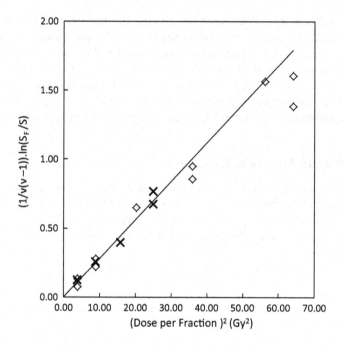

FIGURE 5.17 Cell survival data for fractionated exposures of stationary CHO cells, analysed according to Equation 5.25. The open diamonds are for two fractions, the crosses are for three fractions. The line defines the quadratic coefficient $p\beta_\infty = 0.028$ Gy^{-2}, as found in Figure 5.13. Please note the horizontal axis is the square of the fraction dose (data from Chadwick et al. 1998).

and

$$\left(S/\left(S_F\right)^{\upsilon}\right) = \exp\left(p\upsilon(\upsilon-1)\alpha D_F\right) \tag{5.27}$$

so that

$$(1/\upsilon(\upsilon-1)) \cdot \ln\left(S/\left(S_F\right)^{\upsilon}\right) = p\alpha D_F. \tag{5.28}$$

Equation 5.28 predicts that a plot of $(1/\upsilon(\upsilon-1)) \cdot \ln(S/(S_F)^{\upsilon})$ against (D_F) will give a straight line of slope $(p\alpha)$ for all fraction data which satisfy the restrictions imposed by the analysis (see Figure 5.18).

Only the lower 'dose per fraction' data give a reasonable analysis.

5.4.4 OTHER PUBLISHED DATA

Two related publications by Bedford and Cornforth (Cornforth and Bedford 1987; Bedford and Cornforth 1987) describe a detailed investigation of the cellular effects of fractionated exposures which takes all the experimental precautions needed for a useful analysis. The publications of Bedford and Cornforth considered the

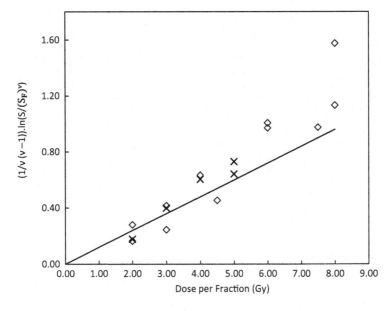

FIGURE 5.18   Cell survival data for fractionated exposures of stationary CHO cells, analysed according to Equation 5.28. The open diamonds are for two fractions, the crosses are for three fractions. The line defines the linear coefficient $p\alpha = 0.12$ Gy^{-1}, as found in Figure 5.13 (data from Chadwick et al. 1998).

relationship between repair of sub-lethal and potentially lethal damage and chromosome rejoining in stationary cultures of primary human fibroblast cells and presented data on survival. Bedford and Cornforth used a 6-hour delay between fractions and a 24-hour delay after the last exposure before plating the cells.

The following figures present the analysis of the data of Bedford and Cornforth on cell survival. Figure 5.19 presents all the cell survival data for acute and fractionated exposures in the form of Equations 5.16 and 5.18.

Figure 5.20 presents their data, analysed according to Equation 5.25, to derive the quadratic coefficient of survival ($p\beta_\infty$).

Figure 5.21 presents their data, analysed according to Equation 5.28, to derive the linear coefficient of survival, ($p\alpha$).

Again, only the lower 'dose per fraction' data give a reasonable analysis.

Contrary to the suggestion by Bedford and Cornforth (1987) that the repair of 'sub-lethal damage' and the repair of 'potentially lethal damage' are two manifestations of the same process, here the 'sub-lethal damage' is identified as DNA single strand breaks and the 'potentially lethal damage' is defined as DNA double strand breaks. Two different processes are involved.

The good agreement found between the theory and the experiment for cell survival, following fractionated exposures, strongly supports the use of the linear–quadratic equation for the analysis of cell survival data. Equations 5.10 to 5.28, developed from the model, provide an all-inclusive analysis of dose fractionation. The effect of fractionation is ascribed to the perfect repair of DNA single strand breaks in between fractions.

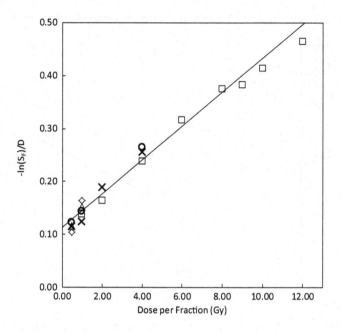

FIGURE 5.19 Cell survival data for stationary primary human fibroblast cells, exposed to acute and fractionated doses analysed according to Equations 5.16 and 5.18. Open squares are acute exposures, open circles represent fractions of 0.5, 1.0 and 4.0 Gy to a total dose of 12 Gy, crosses represent fractions of 0.5, 1.0, 2.0 and 4.0 Gy to a total dose of 8 Gy, diamonds represent fractions of 0.5 and 1.0 Gy to total doses of 4 and 5 Gy. The line has an intercept of $p\alpha = 0.113$ Gy^{-1} and a slope of $p\beta_\infty = 0.032$ Gy^{-2} (data from Cornforth and Bedford 1987).

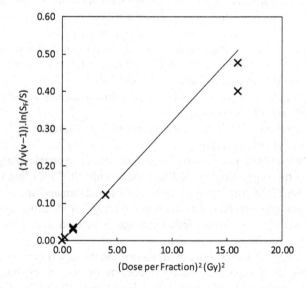

FIGURE 5.20 Fractionated cell survival data for stationary primary human fibroblast cells, analysed according to Equation 5.25, to determine the quadratic coefficient $p\beta_\infty = 0.032$ Gy^{-2}. The crosses represent from 2 to 16 equal fractions (data from Cornforth and Bedford 1987).

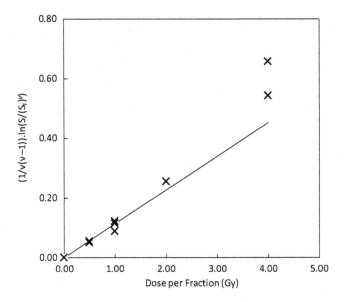

FIGURE 5.21 Fractionated cell survival data for stationary primary human fibroblast cells, analysed according to Equation 5.28, to determine the linear coefficient $p\alpha = 0.113$ Gy^{-1}. The crosses represent data from 2 to 16 equal fractions (data from Cornforth and Bedford 1987).

5.5 THE EFFECT OF POST-IRRADIATION REPAIR

5.5.1 Cell Survival

When stationary but metabolically active cells are plated with a delay after an acute exposure, the cell survival curve shows, in general, an increase in survival compared with the survival when cells are plated immediately after exposure. This is explained in the model by the repair of DNA double strand breaks in the delay period between the end of exposure and the plating of the cells.

This can be understood on the basis of a change in the value of the parameter (f_0) for the repair of DNA double strand breaks. If the value of the repair parameter after immediate plating is (f_{0I}) and the value of the parameter after delayed plating is (f_{0D}), then the number of DNA double strand breaks after immediate plating (N_I) is

$$N_I = f_{0I}\left(\alpha_0 D + \beta_\infty D^2\right) \tag{5.29}$$

where (α_0) and (β_∞) are the coefficients with no post-irradiation repair.

The number of double strand breaks after delayed plating (N_D) is

$$N_D = f_{0D}\left(\alpha_0 D + \beta_\infty D^2\right). \tag{5.30}$$

The change in (f_0) affects the linear (α) coefficient and the quadratic (β) coefficient equally.

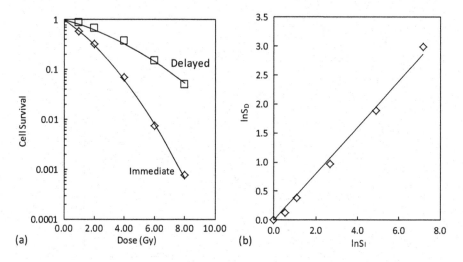

FIGURE 5.22 Cell survival (a) of stationary primary human fibroblasts following an acute exposure for immediate plating and with a delay of 24 hours before plating, showing considerable repair, and (b) the same data analysed according to Equation 5.33 (data from Cornforth and Bedford 1987). The straight line has a slope of 0.4, the immediate plating survival is $S_I = \exp(-(0.475D + 0.055D^2))$ and the delayed plating survival is $S_D = \exp(-(0.19D + 0.022D^2))$. Both the linear and quadratic coefficients differ by the product of 0.4.

Cell survival after immediate plating is

$$S_I = \exp\left(-pf_{01}\left(\alpha_0 D + \beta_\infty D^2\right)\right) \tag{5.31}$$

and after delayed plating is

$$S_D = \exp\left(-pf_{0D}\left(\alpha_0 D + \beta_\infty D^2\right)\right). \tag{5.32}$$

These two survival data are related, so that

$$\ln S_D = \left(f_{0D} / f_{01}\right) \cdot \ln S_I. \tag{5.33}$$

This equation predicts that the plot of ($\ln S_D$) against ($\ln S_I$) should give a straight line with a slope of (f_{0D}/f_{0I}).

Figure 5.22 presents data from Cornforth and Bedford (1987) for survival of stationary primary human fibroblast cells on immediate plating and with a delay in plating of 24 hours after acute exposures, analysed according to Equation 5.33.

5.5.2 CHROMOSOME ABERRATIONS

In analogy with the case of cell survival, it is possible to derive the relationship between the yield of chromosome aberrations on immediate culture and the yield on delayed culture. The yield on immediate culture is

$$Y_I = cf_{0I}\left(\alpha_0 D + B_\infty D^2\right) \tag{5.34}$$

and the yield on delayed culture is

$$Y_D = cf_{0D}\left(\alpha_0 D + \beta_\infty D^2\right). \tag{5.35}$$

These two yields are related, so that

$$Y_D = \left(f_{0D} / f_{0I}\right) Y_I. \tag{5.36}$$

Equation 5.36 predicts that the yield of aberrations after delayed culture will be linearly related to the yield of aberrations after immediate culture.

Nagasawa and Little (1981) reported the decrease in chromosome aberration yield in stationary C3H 10T1/2 mouse cells with time after exposure to a dose of 2 and 4 Gy of X-rays. They also presented data showing that the linear correlation between the logarithm of survival and the yield of chromosomal aberrations found after immediate plating, did not change with time when cells were held for repair after exposure. This is in accordance with Equation 3.4, presented in Chapter 3.

Cornforth and Bedford (1983) also reported that the yield of chromosome aberrations in stationary AG1522 human fibroblast cells decreased with time after exposure to 3.3 and 6 Gy of X-rays. These data are in accordance with the results of Nagasawa and Little (1981) and with the Equations 5.34 and 5.35 but, contrary to the suggestion made by Cornforth and Bedford (1983), their results are not inconsistent with the model of Chadwick and Leenhouts (1973a, 1978), developed and presented here.

5.5.3 MUTATION FREQUENCY

In the same way, using the approximate equation for mutation frequency per surviving cell, the relationship between the mutation frequency on immediate and on delayed plating can be derived as

$$M_D = \left(f_{0D}/f_{0I}\right) M_I. \tag{5.37}$$

Equation 5.37 predicts that the mutation frequency after delayed plating will be linearly related to the mutation frequency after immediate plating.

In Chapter 3, the correlation between survival and mutations was presented from the work of Rao and Hopwood (1982) and Iliakis (1984a,b). In both cases, the same correlation between survival and mutations was made for immediate and for delayed plating. Figure 5.23 presents data from Iliakis (1984a,b) for survival in stationary mammalian cells on immediate and delayed plating with an analysis according to Equation 5.33. Figure 5.24 presents the data from Iliakis for mutations on immediate and delayed plating with an analysis according to Equation 5.37.

The interesting feature of these results is that the increase in survival on delayed plating is exactly matched by the decrease in mutations on delayed plating. The same was found by Rao and Hopwood (1982).

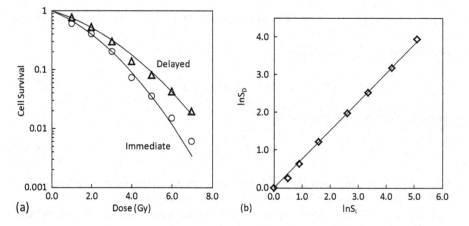

FIGURE 5.23 Cell survival (a) data from Iliakis (1984a,b) after immediate and delayed plating with a plot (b) of the two survival data according to Equation 5.33.

There is an explanation for these results in the model. The explanation lies in the Resnick recombination repair process for DNA double strand breaks. In all the figures shown in Chapter 4 for the formation of chromosomal aberrations from the repair of a DNA double strand break, two different possible resolutions of the repair process were illustrated. If the Holliday hetero-duplex was properly resolved, then the broken double helix was restored. This accounts for both the reduction in chromosomal aberration yield and the mutation frequency. If the Holliday hetero-duplex was incorrectly resolved, then an aberration and, most probably, a mutation would be formed. It seems that delayed plating in a stationary state offers the cell more time to resolve the Holliday hetero-duplex correctly. Immediate plating possibly rushes

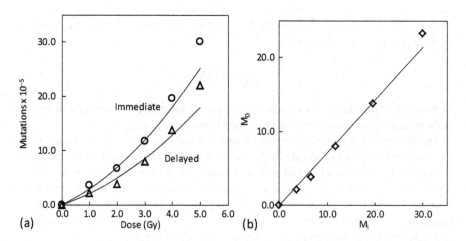

FIGURE 5.24 Mutation data (a) from Iliakis (1984a,b) after immediate and delayed plating illustrating the reduction in mutation frequency on delayed plating. (b) The mutation data, plotted according to Equation 5.37.

the repair of the double strand break so that the cell sometimes fails to resolve the Holliday hetero-duplex properly.

5.6   CONCLUSIONS

The perfect repair of DNA single strand breaks during a protracted or chronic exposure explains the dose rate effect which is revealed in a reduction of the quadratic ($p\beta$) coefficient, ultimately, to zero.

The perfect repair of DNA single strand breaks in-between two fractionated doses explains the restoration of the shoulder of the survival curve and a reduction in effect when a long-enough interval is allowed for repair between the two dose fractions.

The effect of the delayed plating of stationary cells, exposed to acute doses of radiation, has been shown to have a comparable reduced effect on cell survival, chromosomal aberrations and mutations and, in accordance with the model, this defines the post-irradiation repair process as the repair of DNA double strand breaks which is unlikely to be always perfect.

The post-irradiation repair of DNA double strand breaks, defined by the parameter (f_0), applies equally to both the linear (α) coefficient and the quadratic (β) coefficient and should, therefore, be detectable after exposure to both sparsely and densely ionising radiation, even if the dose–effect curves for the different cellular effects have a very small or negligible quadratic (β) coefficient. This is different from the case for the repair of DNA single strand breaks during long exposures or fractionation intervals, which only affects the quadratic (β) coefficient.

Finally, what has been known in the past as 'sub-lethal damage' is identified here as DNA single strand breaks. Its repair affects the quadratic (β) coefficient of the linear–quadratic dose–effect relationship. What has been known in the past as 'potentially lethal damage' is identified here as DNA double strand breaks. Its repair affects both the linear (α) and quadratic (β) coefficients equally.

6 Radiation Quality

The linear coefficient of the dose response for double strand breaks arises from the instantaneous induction of a break in each strand of the DNA during the passage of a single ionising particle. More densely ionising radiations are more efficient at this process and have a higher relative biological effectiveness (RBE). The maximum RBE of a radiation is the ratio of its linear coefficient to the linear coefficient of the test radiation, usually sparsely ionising radiation, and is valid at very low doses. A qualitative and quantitative appreciation of the effectiveness of different radiations is presented. The quantitative appreciation makes use of the three-dimensional structure of the DNA double helix target and a theoretical description of the spatial energy deposition pattern of the radiation track at a nanometre resolution in three dimensions, taking account of all the secondary particles arising in the slowing down of the primary radiation track. The results of several analyses of the variation of the linear coefficient of the dose response for DNA double strand breaks and cell survival, measured after exposure to densely ionising radiations, are presented. The anticipated variation of the linear coefficient for different sparsely ionising radiations is discussed.

6.1 INTRODUCTION

Different radiations have a different biological effectiveness per unit dose and are generally classified under the title of radiation quality into 'sparsely' ionising radiations, such as gamma rays, X-rays, beta radiation and energetic electrons, and 'densely' ionising radiations, such as alpha particles and accelerated ions. It is probably true to say that, in between, there is also a sub-category of 'less sparsely to more densely' ionising radiations, such as soft X-rays, low energy beta radiation, energetic protons and fast neutrons. These different categories are also often identified with their 'linear energy transfer' (LET), a measure of the level of energy deposition along the radiation track in kiloelectron volts per micrometre (keV/µm). Sparsely ionising radiations have 'low LET' and densely ionising radiations have 'high LET'. The LET concept will not be used here because, although it provides a quantitative parameter for the description of a radiation type, it does not provide a unique identification of the biological effectiveness of the radiation. Different radiations with the same LET can have a differing biological effectiveness (Barendsen 1964; Barendsen et al. 1966; Todd 1967).

The dose–effect relationships for cellular endpoints measured for different radiations have different shapes as well as different relative effectiveness. For example, acute exposure to sparsely ionising radiations results in linear–quadratic dose–effect curves but acute exposure to densely ionising radiations results in linear dose–effect curves (see Figure 6.1). Clearly, the pattern of energy deposition by the radiation affects the shape of the dose–effect curves and the radiation's biological

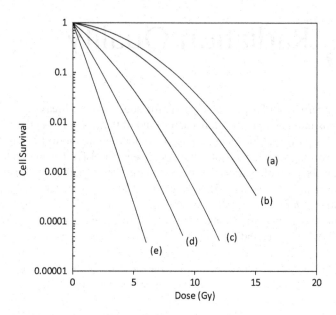

FIGURE 6.1   The changing shape of cell survival curves for acute exposure to different types of radiation, from sparsely ionising radiations to densely ionising radiations. (a) Gamma rays, (b) 250 kVp X-rays, (c) 0.3 keV ultrasoft X-rays, (d) alpha particle (He ion), (e) carbon ion.

effectiveness. This implies that the biological target has some structure, must require more than one energy deposition event to cause the cellular effect and that the energy deposition events must be spatially related.

The DNA double helix is a well-defined molecular three-dimensional structure with nanometre dimensions. A brief examination of Figure 2.14 and Equation 1.8 for the induction of DNA double strand breaks, that is,

$$N = \left(\alpha D + \beta D^2\right) = \left(2f_0 n\mu k\Omega kD + f_0 f_1 n\mu k\left(1 - \Omega k\right)n_1\mu_1 k_1 D^2\right), \qquad (6.1)$$

reveals that the induction of DNA double strand breaks in the passage of one ionising particle track (alpha mode, linear (α) coefficient) involves two instantaneous energy depositions, spatially related so that one energy deposition occurs close to each strand of the double helix. Consequently, the pattern of energy deposition events at the nanometre scale along the radiation track is primarily involved in the induction of DNA double strand breaks in the alpha mode and affects the linear (α) coefficient of Equation 6.1. Detailed examination of the different coefficients included in the linear–quadratic Equation 6.1 identifies the parameter group ($\mu k\Omega k$), which defines the probability that an ionising radiation track passes close to one DNA strand and there is an energy deposition close to the strand causing a break in that strand, and that the track continues to pass close to the second DNA strand and that there is a second energy deposition close to the second strand causing a break in that strand (see Chapter 1). This probability depends on the spatial deposition of energy events along and around the ionising radiation track in relation to the molecular dimensions of the DNA double helix. The linear (α) coefficient in the induction of DNA double strand breaks is thus closely dependent on, and is defined by, the radiation quality. In general, sparsely ionising radiation has a smaller linear (α) coefficient compared with that of densely ionising radiation.

A detailed examination of the quadratic coefficient in Equation 6.1 identifies the parameter group $(1 - \Omega k)$ which defines the total probability, per 'first' strand break, that the 'second' strand is not broken in the passage of the same ionising particle. Clearly, sparsely ionising radiation is not an efficient inducer of DNA double strand breaks in the passage of a single ionising track and $(1 - \Omega k)$ is close to unity. Densely ionising radiation is more efficient at inducing DNA double strand breaks in the passage of a single track; Ωk increases in value and approaches the value of 1. This means that $(1 - \Omega k)$ becomes very small and, consequently, the quadratic (β) coefficient becomes smaller and the dose–effect curve for the induction of double strand breaks is defined predominantly by the linear (α) coefficient (see survival curves in Figure 6.1). This also means that densely ionising radiation effects will exhibit little, if any, dose rate effect which is only expressed in the quadratic (β) coefficient and there will be little or no effect of dose fractionation.

Thus, the increase in radiation quality from sparsely ionising radiation to densely ionising radiation is reflected in an increase in the value of the linear (α) coefficient. However, in contrast to the case of decreasing dose rate which leads to a decrease in the quadratic (β) coefficient but does not affect the linear (α) coefficient (see Chapter 5), the increase in the linear (α) coefficient caused by densely ionising radiation does have a decreasing effect on the quadratic (β) coefficient.

6.2 RELATIVE BIOLOGICAL EFFECTIVENESS

Traditionally, the differing biological efficiencies of different radiations have been measured by defining the relative biological effectiveness (RBE) as the ratio of the dose of a standard radiation (often 250 kVp X-rays) to the dose of the test radiation which caused the same biological effect (see Figure 6.2):

$$RBE = Dose\,(standard\ radiation)\,/\,Dose\,(test\ radiation). \qquad (6.2)$$

The RBEs are listed below.

	RBE_1	RBE_{10}	RBE_0
Gamma rays	0.9	0.85	0.46
0.3 keV X-rays	1.6	1.8	3.4
He ion	2.4	3.0	6.5
Carbon ion	3.9	5.0	11.0

This reveals that the value of the RBE depends on the effect level (1% and 10% survival) at which the RBE is determined. This is a result of the varying shapes of the linear–quadratic dose–effect relationships for the different radiations. Also included in the list is the RBE_0, which is the 'limiting relative biological effectiveness' at very

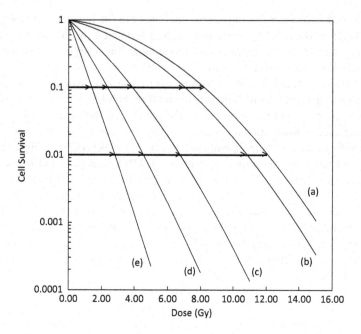

FIGURE 6.2 A representation of Figure 6.1 with arrows indicating the doses for each radiation giving 1% and 10% cell survival. The RBEs for each radiation can be determined at both cell survival levels from these doses using 250 kVp X-rays (b) as standard radiation.

low doses. It has been defined as the ratio of the linear (α) coefficient of the test radiation to the linear (α) coefficient of the standard radiation (Chadwick and Leenhouts 1981), that is,

$$\text{RBE}_0 = \alpha \text{ value test radiation} / \alpha \text{ value standard radiation} = \alpha_t / \alpha_s. \qquad (6.3)$$

The advantages of RBE_0 are that it is a constant value, it is valid at low doses and it is independent of dose rate and, consequently, it is relevant for the weighting factor of different radiations used in radiological protection. The weighting factor is used in radiological protection to take the differing biological effectiveness of different radiations into account. The dose of radiation in Gray (Gy) is multiplied by the weighting factor to create the dose equivalent in Sievert (Sv). This makes it possible to add the exposures of different radiations together in a single unit, the Sievert. The RBE_0 is valid for DNA double strand breaks and each of the three cellular endpoints – survival, chromosome aberrations and mutations – and, by extension, to the induction of cancer and hereditary mutations. Importantly, it can be determined using cellular radiobiological measurements.

The list reveals that for radiations more densely ionising than the standard radiation the RBE_0 is greater than unity and is a maximum of the RBE values measured at different levels of biological effect. For very sparsely ionising radiation, such as cobalt-60 gamma rays, the RBE_0 is less than 1 at 0.46 and is the minimum value of the RBE values measured at different levels of biological effect. This value of 0.46

for cobalt-60 gamma rays, measured against 250 kVp X-rays, signals that the RBE_0 for all sparsely ionising radiation is not 1. This contradicts the choice of the value of 1 for the weighting factor of all sparsely ionising radiation currently used in radiological protection.

It has to be noted that although the RBE_0 for a given test radiation is a constant, its value can depend on the experimental conditions under which it is determined. The RBE_0 is unlikely to vary much for sparsely ionising radiations measured under different experimental conditions, for example, aerobic or anaerobic, but it could vary somewhat for more densely ionising radiations. An example of this is revealed by comparing the variation of the linear (α) coefficient through the cell cycle shown in Chapter 2, Figure 2.9, where the decrease in the (α) coefficient in the S phase is relatively larger for the sparsely ionising X-rays (test radiation) than for the more densely ionising neutron and alpha particle radiations. The data in Figure 2.9 give a variation of RBE_0 for the neutrons from 4 in G_1 phase to 7 in S phase but this is essentially because of the more pronounced change in the linear (α) coefficient in the sparsely ionising test radiation.

Although the 'limiting relative biological effectiveness (RBE_0)' is a much more useful measure than the older 'relative biological effectiveness', it still has problems with respect to its use for the definition of weighting factors in radiological protection. This is because different sparsely ionising radiations will exhibit a considerable deviation from unity, the weighting factor currently used by regulatory authorities in recommendations for radiological protection and which is used as a base to derive weighting factors for other more densely ionising radiations.

6.3   A QUALITATIVE APPRECIATION OF THE EFFECT OF RADIATION QUALITY

Figure 6.3 presents a series of schematic representations of different types of radiations passing close to the DNA double helix, which should permit an intuitive understanding of why different radiations have different efficiencies in inducing DNA double strand breaks in the passage of a single ionising track (the alpha mode). The particle tracks are represented by the pale grey cylinders and the energy depositions are represented by the small stars. The first image in Figure 6.3 shows (a) an energetic electron track scattered from energetic gamma rays and X-rays, then (b) a soft low energy electron track, (c) a proton track, (d) an alpha particle track and, finally, (e) an accelerated ion track.

Figure 6.3a shows an energetic electron track with a narrow track core (pale grey cylinder) with energy depositions (white stars) every 5–7 nm on average, which will not be very effective. Figure 6.3b shows a much less energetic electron track with energy depositions every 1–2 nm on average, which will be relatively much more effective. Figure 6.3c shows an energetic proton track with a wider track core (pale grey cylinder) with energy depositions every 1–2 nm on average, which will be even more effective because it has a better chance of passing close to both strands of the helix than the narrow electron tracks. Figure 6.3d shows an alpha particle track with a wide track core and energy depositions every 1–2 nm on average, which will be very efficient. Figure 6.3e shows an accelerated heavy ion with a wide track core

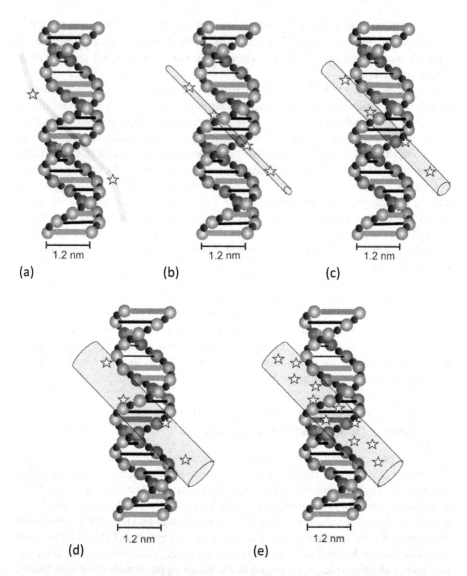

FIGURE 6.3 Schematic diagrams representing different tracks passing close to the DNA double helix to give an intuitive feel for the changing effectiveness of different radiations in inducing a double strand break in the alpha mode of radiation action.

but with energy depositions every 0.5 nm which will be less effective than the alpha particle track because it deposits more energy within the DNA double helix than is necessary (4–5 energy depositions instead of 2), a situation known as 'overkill'.

Clearly, the radiation tracks with energy depositions every 1 to 2 nanometres are the most efficient and the tracks with a large track radius will be more efficient than those with a small track radius because, if they pass close to the DNA double helix, they will have a better chance of passing close to both strands. In this respect,

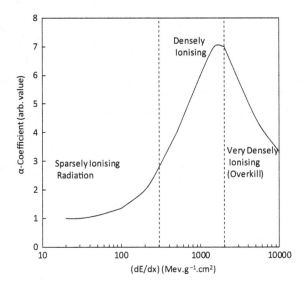

FIGURE 6.4 The anticipated variation in the linear (α) coefficient with radiation quality on the basis of Figure 6.3.

it is worth noting that Watt and colleagues (Watt and Kadiri 1990; Watt et al. 1985) presented analyses which suggested that the highest biological efficiency was found for different radiations that had energy depositions every 1 to 2 nanometres, and that Prise and colleagues (Prise et al. 1993, 1999) showed that DNA double strand breaks arose from paired radical sites caused by two energy deposition events.

Figure 6.3a and b provide an intuitive appreciation of why the 'limiting relative biological effectiveness (RBE$_0$)' varies for different sparsely ionising radiations. The energetic secondary electrons arising from gamma rays and X-rays shown in Figure 6.3a will be very inefficient at inducing DNA double strand breaks in the passage of a single track and it will only be after attenuation, at the end of the low energy secondary electron tracks, that these very soft electrons become efficient inducers of double strand breaks, as shown in Figure 6.3b. The biological effectiveness of a sparsely ionising radiation will, consequently, depend on the proportion of radiation dose deposited by the efficient low energy soft electrons compared to the total dose.

The intuitive understanding of the implications of Figure 6.3 leads to a generalised expected variation of the linear (α) coefficient as a function of the probable energy deposition pattern of different radiations close to the DNA double helix, as shown in Figure 6.4. For comparison, Figure 6.5 presents the induction of DNA double strand breaks in ΦX-174 phage as a function of ion stopping power (Christensen et al. 1972) and Figure 6.6 presents the induction of DNA double strand breaks, measured as unrepaired single strand breaks, in Chinese hamster cells as a function of ion stopping power (Ritter et al. 1977).

There is clear evidence that the actual dependence of the induction of DNA double strand breaks on radiation quality matches the anticipated qualitative dependence derived from a consideration of Figure 6.3.

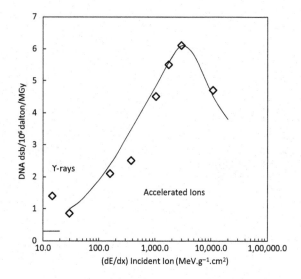

FIGURE 6.5 The induction of DNA double strand breaks in ΦX-174 phage in radio-protective broth by heavy ions as a function of the ion stopping power (Christensen et al., 1972). (The values of (dE/dx) do not apply to the gamma ray data point).

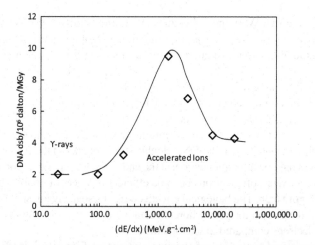

FIGURE 6.6 The heavy ion induction of DNA double strand breaks, measured as unrepaired single strand breaks, in cultured Chinese hamster cells as a function of ion stopping power (from Ritter et al. 1977). (The values of stopping power do not apply to the gamma ray data point.)

Figure 6.7 shows, in contrast, the induction of single strand breaks in Chinese hamster cells as a function of ion stopping power (Ritter et al. 1977) which illustrates the expected decrease in efficiency as the stopping power increases (see Section 6.1). Figure 6.3 also identifies the importance of the spatial distribution of energy depositions along a particle track in relation to the DNA double helix and the concept of a track core of nanometre dimensions. Both of these factors are crucial to the quantitative calculation of the dependence of biological effectiveness on radiation quality, as outlined in the following section.

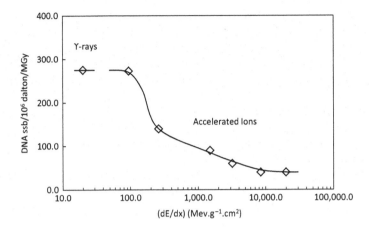

FIGURE 6.7 The heavy ion induction of DNA single strand breaks in Chinese hamster cells as a function of ion stopping power (from Ritter et al. 1977). (The values of stopping power do not apply to the gamma ray data point.)

6.4 A QUANTITATIVE APPRECIATION OF THE EFFECT OF RADIATION QUALITY

The well-defined three-dimensional structure of the DNA double helix with nanometre dimensions means that any consideration of the interaction of radiation tracks with the DNA double helix in the production of DNA double strand breaks depends on a knowledge of the spatial energy deposition pattern of the radiation track at a nanometre resolution in three dimensions. It has been possible to develop a theoretical approach to this problem from first principles by developing a track model for interaction with water, assuming that the same processes and quantities will apply in the cell, as biological material is more than 70% water.

The track model has to be able to deal with the primary track, each subsequent δ-ray track and each electron scattering event, independently. An ionising particle loses energy as it travels through matter by transferring energy to the atomic electrons of the matter. The scattered energetic electrons are ejected from the atoms in ionisation events and lose their energy as secondary electrons by also scattering the atomic electrons of the matter. This continues until the energy of the electrons is so low that they are no longer able to eject an atomic electron and cause an ionisation event. As the ionising particle loses energy, it creates an avalanche of secondary electrons of decreasing energy until further ionisation ceases.

The curious reader is referred to *The Molecular Theory of Radiation Biology* (Chadwick and Leenhouts 1981), where a complete derivation of the theoretical radiation track model and its interaction with the DNA double helix is presented in Chapter 8 of the book. The radiation track model was subject to several developments to describe the spatial distribution of energy deposition events at the nanometre scale and also to examine its usefulness in predicting the effects in a single hit detector, such as the Fricke dosimeter, and in the induction of DNA double strand breaks. (Leenhouts and Chadwick 1974a, 1976, 1985; Pruppers et al. 1990; Leenhouts et al. 1990).

In the following, a brief description of the track model and the processes involved in the interaction of the track with the DNA is presented together with some results of the application of the model to the analysis of experimental data.

The 'slowing down' of an ionising particle in water is described by its stopping power (–dE/dx). The stopping power depends on the electron density in water (N_{el}), the scattering probability per electron (P_{sc}), the mean energy lost per scattering event (Q_{av}) and the effective radius (R) of the track core within which the primary electron scattering events occur. The track core is cylindrical because the elastic electron scattering is isotropic with respect to the particle track path.

The stopping power (–dE/dx) can be written as

$$-dE/dx = \pi R^2 N_{el} \cdot P_{sc} \cdot Q_{av}. \tag{6.4}$$

The electron density in water (N_{el}) is 3.3×10^{23} electrons/g. The effective radius (R) of the primary track includes the maximum impact parameter, the uncertainty in (R) due to the Heisenberg uncertainty principle and the molecular dimensions of an ion which has captured an electron. The scattering probability is derived from the Rutherford scattering cross-section and the mean energy (Q_{av}) lost per scattering event is derived from the excitation spectrum of water and the maximum energy transfer in a head-on collision using the Platzman excitation spectrum of water (Platzman 1967).

A series of assumptions have been made in the calculation of DNA single and double strand breaks, namely:

- The DNA molecule is visualised as a linear array of DNA bases every 0.34 nm, each linked to a second linear array of bases at 1.2 nm distance.
- Strand breakage occurs via radiolysis products (radicals) produced in the water surrounding the DNA molecule.
- The radicals have a limited lifetime and thus a limited diffusion range.
- A water molecule is ionised or excited when one of its 10 electrons is scattered by the charged particle.
- Ionisation involves energy transfers of more than 20 eV, excitation involves energy transfers between 5 and 20 eV and energy transfers of less than 5 eV are not effective.
- F_{ion} is the number of effective radiolysis products per water molecule created by ionisation events and F_{exc} is the number of radiolysis products created by excitation events.
- To be effective, the radiolysis products have to be created within their diffusion range (ρ) of the DNA.

The equation for DNA double strand break induction in the passage of one particle track without repair, the linear (α) coefficient, is given in Equations 1.2 and 6.1 as

$$\alpha_0 = 2n\mu k\Omega k. \tag{6.5}$$

And equally, the number of single strand breaks (N_{ssb}) induced in a DNA molecule per unit dose, which is the number of 'first' strand breaks when the track misses and does not break the 'second' strand, is given by

$$N_{ssb} = 2n\mu k \left(1 - \Omega k\right) \tag{6.6}$$

where

    n   is the number of nucleotide base pairs per cell.

    μ is the probability per unit dose that an ionising particle passes close to a nucleotide base. The value of μ includes the effective particle track radius (R), the effective diffusion range of the radiolysis products (ρ) and the stopping power ($-dE/dx$).

    k   is the probability per nucleotide base that, when the particle passes close to the DNA, an energy deposition occurs which leads to a strand break.

    Ω is the probability that, when the ionising particle passes close to the 'first' DNA strand, it also passes close to the 'second' strand.

An expression for μ can be derived by defining a 'unit volume of interaction per nucleotide base' as the cylinder of interaction around the DNA strand having a radius of $(R^2 + \rho^2)^{1/2}$ with a height of 0.34 nm (see Figure 6.8). Any particle passing through the cylinder can produce an effective radiolysis product. This leads to an 'apparent interaction cross-section $(A = 2(R^2 + \rho^2)^{1/2}(0.34))$' which is a measure of the area of the unit volume of interaction seen by a particle passing through the cylinder. As the reciprocal of the stopping power is the probability per unit area per unit dose that an ionising particle crosses the area (A), μ can be defined as:

$$\mu = A \cdot \left(-dx/dE\right) = 2\left(R^2 + \rho^2\right)^{1/2} \cdot 0.34 \text{ nm} \cdot \left(-dx/dE\right). \tag{6.7}$$

An expression for k can be derived by defining a 'cylinder of migration' around a DNA strand with a radius of (ρ), the diffusion distance of a water radical. It is assumed that any water radical induced in this cylinder of migration will cause a

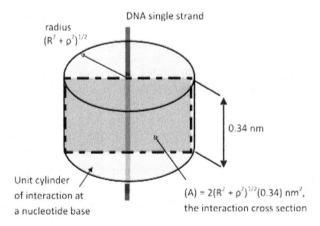

FIGURE 6.8 A schematic representation of a 'unit cylinder of interaction' (pale grey) round a DNA single strand at a nucleotide base. When an ionising particle track passes through any part of this cylinder, it can create an effective hydrolysis product leading to a break in the strand. Also shown is the apparent interaction cross-section (A) (dark grey area).

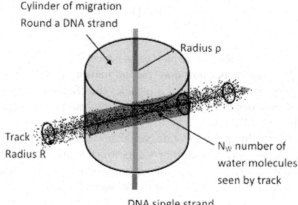

Cylinder of migration
Round a DNA strand

Radius ρ

Track
Radius R

N_w number of
water molecules
seen by track

DNA single strand

FIGURE 6.9 A schematic drawing showing a 'unit cylinder of migration' (grey), radius (ρ), round a DNA single strand at a nucleotide base. Each hydrolysis product produced in the cylinder is assumed to cause a strand break. Also shown is the volume which is common to the cylinder of migration and the track core (dark grey).

strand break. The probability that a water radical is formed within this cylinder of migration is derived by multiplying the number (N_w) of water molecules in the common volume of the cylinder of migration, radius (ρ), and a particle track of radius (R) crossing that cylinder (see Figure 6.9) by the probability (P_w) that one or more of the 10 electrons in a water molecule is scattered and the number (F) of water radicals induced by the scattering event. (N_w) is a complicated function of (ρ) and (R).

$$P_w = 1 - \exp(-10 P_{sc})$$
(6.8)

and

$$F = \Delta F_{exc} + (1 - \Delta) F_{ion}$$
(6.9)

where (Δ) is the fraction of scattering events giving excitation and depends on the maximum energy transfer of the ionising particle.

k is thus

$$k = 1 - \exp(-N_w \cdot F(1 - \exp(-10 P_{sc}))).$$
(6.10)

The parameter Ω depends on the radius of the track (R), the diffusion range of the water radical (ρ) and the distance of 1.2 nm between the two strands of the DNA double helix. Ω is the probability that an ionising particle track of radius (R) passes through two 'cylinders of migration' of radius (ρ) which have their central axes separated by 1.2 nm (see Figure 6.10a).

Using this definition leads to an approximation for Ω based on the probability that two 'sites', separated by 1.2 nm, both fall within a cylinder around the particle track with a radius of $(R^2 + \rho^2)^{1/2}$ (see Figure 6.10b).

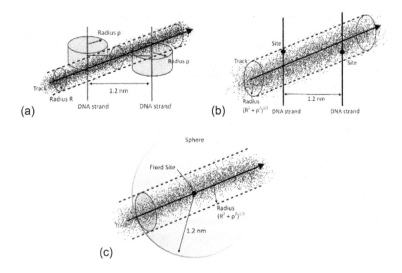

FIGURE 6.10 (a) A schematic drawing of two 'cylinders of migration', radius (ρ), round each DNA strand of a double helix (separation 1.2 nm) and an ionising particle track, radius (R), crossing both cylinders. (b) A schematic drawing of two 'sites' on the DNA strands, separated by 1.2 nm and crossed by a track in a 'cylinder of interaction', radius of $(R^2 + \rho^2)^{1/2}$. This is an approximation of the situation shown in (a). (c) A schematic drawing of a 'track', radius $(R^2 + \rho^2)^{1/2}$, crossing a 'fixed site' at the centre of a sphere (pale grey) of radius 1.2 nm. The second site can be situated anywhere on the surface of the sphere. The probability (Ω_i) that the track crosses both 'sites' is given by the two areas of the surface of the sphere intersected by the track (ellipses) divided by the total surface area of the sphere.

If the first 'site' crossed by the 'cylinder of interaction' round the track is fixed, the second 'site' is situated somewhere on the surface of a sphere of radius 1.2 nm centred on the first 'site' (see Figure 6.10c).

If the first 'site' is fixed somewhere in the cylinder of interaction around the track, then the probability (Ω_i) that the second 'site' is also in the cylinder of interaction around the track is defined by the fraction of the surface of the sphere that is intersected by the cylinder of interaction round the track divided by the total surface of the sphere. The total probability (Ω) is the average value of (Ω_i) for all the possible positions of the first 'site' being in a cross-section of the cylinder of interaction around the track. A light beam experiment was used to simulate the cylinder of interaction around the track to determine the dependence of (Ω) on the ratio $(R^2 + \rho^2)^{1/2}/1.2$ nm (see Figure 6.11). The extreme values were approximated by mathematical functions (Leenhouts and Chadwick 1974a). Figure 6.11 also shows the value of $(1 - \Omega)$ which is relevant for the quadratic (β) coefficient.

Using two normalisation factors derived by fitting the stopping power equation for electrons and protons to data published by ICRU (1970) (see Chadwick and Leenhouts 1981) it is possible to calculate Equation 6.5 for double strand breaks and Equation 6.6 for single stand breaks from first principles with only three variables, namely, the number of effective water radicals per excitation (F_{exc}), the number of effective water radicals per ionisation (F_{ion}) and the diffusion range of effective water radicals (ρ), which is the radius of the cylinder of migration.

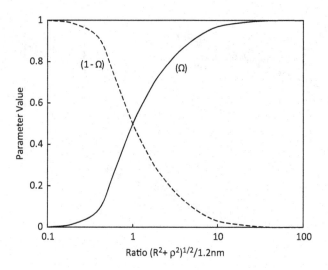

FIGURE 6.11 The parameter (Ω) as a function of the ratio $(R^2 + \rho^2)^{1/2}/1.2$ nm.

However, Equations 6.5 and 6.6 apply to an ionising particle of a fixed energy, a fixed stopping power and a fixed track radius, but any ionising radiation is made up of a spectrum of different ionising particles, mainly electrons, with a range of different energies, stopping powers and track radii. The complete calculation of Equations 6.5 and 6.6 for the induction of double and single strand breaks, respectively, requires the average on a dose base over all primary and secondary radiation tracks arising from the interaction of the primary track with water, using a complete 'slowing down spectrum' (see Leenhouts and Chadwick 1985, Chadwick and Leenhouts 1981).

Calculations for X and gamma radiation make use of the initial distribution of ionising electrons given by Cormack and Johns (1952) and Burch (1957) or the initial photon spectra from Burke and Petit (1960) and Johns (1969) with appropriate cross-sections for the photo-electric and Compton effects.

An analysis from first principles has been used to estimate the number of DNA double strand breaks measured in ΦX-174 phage irradiated *in vitro*, in radio-protective broth, after exposure to gamma rays and different accelerated ions (Christensen et al. 1972) and the results are shown as the lines in Figure 6.5 for ($F_{exc} = 0$); ($F_{ion} = 0.01$); ($\rho = 7$ nm) with a normalisation factor of 2.

In the case of cell survival, an additional normalisation factor (p) is introduced in the calculations from first principles because cell survival is

$$S = \exp\left(-p\left(\alpha D + \beta D^2\right)\right) \tag{6.11}$$

and the linear coefficient is ($p\alpha = p2f_0 n\mu k\Omega k$).

This means that the values of the linear ($p\alpha$) coefficient derived from experimental survival curves can be calculated from first principles using the three parameters, (F_{exc}), (F_{ion}) and (ρ), plus the normalisation factor (p).

In the introduction to this chapter, it was noted that the concept of linear energy transfer (LET) would not be used because the results of Barendsen (Barendsen 1964; Barendsen et al. 1966) and Todd (1967) had revealed that different radiations having the same LET proved to have a different biological effectiveness. These results put the LET concept into doubt. Barendsen and Todd measured cell survival in the same cell type (T-1 kidney cells) after exposure to radiations with a range of different stopping powers but Barendsen used alpha particles and deuterons of different energies while Todd used nine different accelerated ions having the same velocity but different energies.

Analysis of the results of Barendsen and Todd, using the theoretical calculations from first principles, provided fits to both sets of data using exactly the same three parameters and normalisation factor, namely ($F_{exc}=0.2$); ($F_{ion}=0.4$); ($\rho=0.7$ nm) and the normalisation factor ($p=0.125$) when the number of base pairs per cell is set at $n=3\times10^9$. The unification of these results in the analysis reveals the validity of the theoretical approach, in spite of its limitations, together with the power of using both the DNA molecule as the radiation target and the double strand break as the critical radiobiological lesion. The differences in the biological effectiveness of radiations having the same LET in the results of Barendsen and Todd are explained solely by the different track core radii (R) of the primary radiations and their secondary, scattered ionising particles. This means that the stopping power (or LET) of a radiation is not a unique indicator of that radiation's biological effectiveness. The results of the combined analysis are presented in Figure 6.12, clearly showing that the peak of the linear ($p\alpha$) coefficient occurs at different values of stopping power in the two sets of results.

The diffusion distance of ($\rho=0.7$ nm) of the water hydrolysis product suggests that it is a hydroxyl radical (OH•) (Ward and Kuo 1973; Chapman et al. 1975b), in line with the suggestion of Powers (1974) that the (OH•) radical must be produced within a few water molecule diameters of the DNA double helix for it to be effective. This result is also in agreement with the radical scavenging measurements of Chapman et al. (1975a, 1976) which showed that 'indirect action' does have an important effect on the linear ($p\alpha$) coefficient of cell survival.

The radical scavenging experiments of Chapman et al. (1975a, 1976) also indicated that the linear ($p\alpha$) coefficient and the quadratic ($p\beta$) coefficient of accurate cell survival studies responded differently, indicating that different radio-chemical species were involved in the two mechanisms of radiation action. This reflects the definition in the derivation of the quadratic (β) coefficient, presented in Chapter 1, Section 1.5 and Equation 1.4, of two different parameter groups (μk) and ($\mu_1 k_1$) for the probability that a particle track passes close to a DNA base and deposits energy causing a single strand break in the DNA strand, (μk) for the 'first' single strand break and ($\mu_1 k_1$) for the 'second' single strand break. A distinction was made between the (μk) parameters involved in the induction of the 'first' breaks and the ($\mu_1 k_1$) parameters involved in the induction of the 'second' breaks because it was not certain, when the model equations were initially developed, that the same mechanism (radical species) was involved in this 'second' mode of radiation action. Thus, the quadratic (β) coefficient for the induction of DNA double strand breaks in the passage of two independent particle tracks without repair, is given in Equations 1.4 and 6.1 as

$$\beta = f_1 n \mu k \left(1 - \Omega k\right) n_1 \mu_1 k_1. \tag{6.12}$$

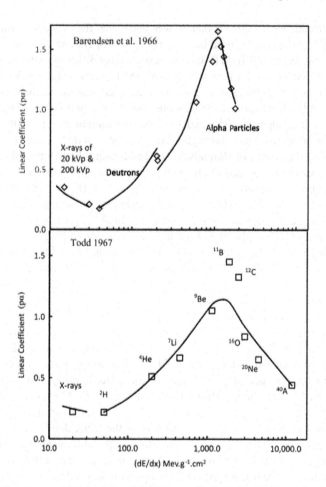

FIGURE 6.12 A comparison of the T-1 kidney cell survival results of Barendsen et al. (1966) and Todd (1967) for the linear (pα) coefficient as a function of stopping power, presenting the data (open diamonds and open squares) for the different radiations used and the theoretical calculations (lines) made from first principles. The theoretical analysis gives $(F_{exc} = 0.2)$; $(F_{ion} = 0.4)$; $(\rho = 0.7$ nm) and $(p = 0.125)$ in both cases. (The values of (dE/dx) do not apply to the X-ray data points.)

It became possible to address this question of different mechanisms in the alpha mode and beta mode of double strand break induction when Roux (1974) published accurate measurements of the survival of *Chlorella* cells for different types of radiation. Roux analysed the survival curves using the linear–quadratic dose relationship to derive values for (pα) and (pβ) for the different radiations used. These values are shown in Figures 6.13 and 6.14 as a function of radiation quality, analysed from first principles, assuming that the DNA content of *Chlorella* cells is about 50 times smaller than the T-1 kidney cells.

The calculated fitting of the linear (pα) coefficients was made with the values $(F_{exc} = 1.0)$; $(F_{ion} = 2.0)$; $(\rho = 0.35$ nm) and $(p = 1)$.

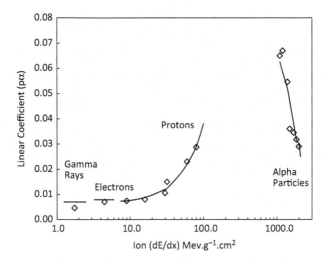

FIGURE 6.13 The fitting, calculated from first principles, of the linear (pα) coefficient derived from survival curves of aerobic *Chlorella* cells (Roux 1974) after exposure to different radiations. (The values of (dE/dx) do not apply to the gamma ray and electron data points.)

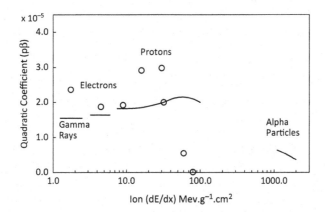

FIGURE 6.14 The fitting, calculated from first principles, of the quadratic (pβ) coefficient derived from survival curves of aerobic *Chlorella* cells (Roux 1974) after exposure to different radiations. (The values of (dE/dx) do not apply to the gamma ray and electron data points.)

The calculated fitting of the quadratic (pβ) coefficients was made with the values $(F_{exc}=1.0)$; $(F_{ion}=2.0)$; $(\rho=0.35\ nm)$ for the 'first' single strand break and $(F_{exc}=1.0)$; $(F_{ion}=2.0)$, $(\rho_1=8.5\ nm)$ for the 'second' single strand break and $(p=1)$.

This indicates that a different radiolysis product is responsible for the induction of the 'second' single strand break which converts the 'first' single strand break into a double strand break. This means that $(\mu_1 k_1)$ does not equal (μk). It was not possible to get an acceptable fitting for the quadratic (pβ) coefficient with $(\mu_1 k_1)$ equal to (μk).

This important result provided a rational quantitative input to support the quadratic (pβ) coefficient in the linear–quadratic dose response equation derived for the

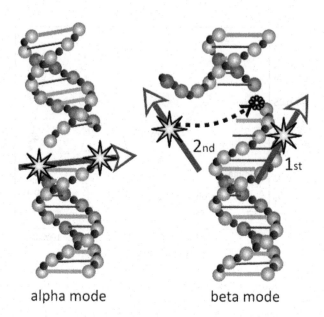

FIGURE 6.15   A schematic representation of the proposed mechanisms involved in the induction of DNA double strand breaks by ionising radiation. In the alpha mode, two hydroxyl radicals are created close to each strand in the passage of a single particle track. In the beta mode, the 'first' single strand break is created by a hydroxyl radical close to one strand and a 'second' active species, created some distance away from the unbroken DNA single strand, converts the 'first' single strand break into a double strand break.

induction of DNA double strand breaks. It is qualitatively in agreement with the results of radical scavenging measurements by Chapman et al. (1975a, 1976) who found different scavenging effects on the linear ($p\alpha$) coefficient compared with the quadratic ($p\beta$) coefficient. These results led directly to the proposal of two different mechanisms for the induction of the two independent single strand breaks involved in the beta mode of radiation action and, as discussed in Chapter 2, Sections 2.4 and 2.5, provided the basis for the redrawing of the schematic representation of the induction of DNA double strand breaks in the alpha mode and beta mode of radiation action, presented in Figure 2.14, and reproduced here in Figure 6.15. This also answered a serious criticism of the model that suggested that, if the same radical processes were involved in the beta mode as in the alpha mode, the two independently induced single strand breaks would be thousands of base pairs apart on opposite DNA strands, a highly unlikely scenario (Kellerer 1975). Indeed, it seems doubtful that the radiolysis product involved in the induction of the 'second' single strand break in the beta mode of radiation action is capable of inducing a single strand break in an intact DNA double helix.

6.5   THE BIOLOGICAL EFFECTIVENESS OF SPARSELY IONISING RADIATIONS

The calculations of the effectiveness of different radiations from first principles, based on the simple radiation model and the concept of a DNA double strand break induced in the alpha mode of radiation action, clearly identify the efficiency of densely ionising radiations such as protons, alpha particles and accelerated ions. These particles are very biologically effective and act by producing two hydroxyl radicals close to each DNA strand in the passage of a single ionising particle track.

However, the calculations from first principles and the concept of a DNA double strand break formed in the passage of a single ionising particle also permit the biological effectiveness of sparsely ionising radiations, such as gamma rays, X-rays and energetic electrons to be taken into account. Qualitative considerations, based on Figure 6.3a and b, reveal that energetic sparsely ionising radiations with energy deposition events every several nanometres along a tortuous path are unlikely to be very effective in inducing a DNA double strand break in the alpha mode of radiation action. All primary sparsely ionising radiations lose energy in a scattering process, producing secondary and subsequent generation energetic electrons which gradually become less energetic in the 'slowing down' process. These low energetic electrons can have energy depositions every 1 to 2 nanometres along their track and can be efficient inducers of DNA double strand breaks in the alpha mode of radiation action. It is, therefore, the proportional dose contribution of these efficient 'soft' electrons to the total dose contribution of the primary and all secondary and subsequent generation electrons of the initial sparsely ionising radiation, which determines the biological effectiveness of the initial radiation. Energetic gamma rays will deposit a considerable amount of ineffective dose with a small contribution of the biologically effective 'soft' electrons. Energetic X-rays and electrons will be similar but, as the energy of the primary X-rays and electrons decreases, the relative dose contribution of the biologically effective 'soft' secondary electrons will increase and the lower energy X-rays and electron primary radiations will be more biologically effective than the higher energetic X-rays and electrons.

This reasoning is strengthened by the work of Goodhead and his colleagues who revealed the high biological effectiveness of very low energy ultrasoft X-rays of 1.5 keV and 0.28 keV in producing all three cellular biological effects (Goodhead et al. 1980; Leenhouts and Chadwick 1985).

There are many examples of experimental measurements of cellular endpoints which reveal the varying biological effectiveness for different sparsely ionising radiations at low doses giving around a fourfold variation in the linear (pα) coefficients (Schmid et al. 1974; Lloyd et al. 1975, 1986; Underbrink et al. 1976; Virsik et al. 1977; Frankenberg et al. 2002; Heyes and Mill 2004; Heyes et al. 2006, 2009; Fairlie 2007; Bridges 2008; Goodhead 2009; Harrison 2009). Most of these publications express concern for the problems associated with this variation, compared with the use of a weighting factor of 1 in radiological protection, for all sparsely ionising radiations. The identification of the DNA double strand break as the critical radiobiological lesion, together with the development of the radiation model for calculations from first principles, lead inescapably to the expectation that different sparsely ionising radiations will have a different biological effectiveness in the alpha mode of radiation action and, thus, provide a theoretical and mechanistic basis for these publications.

This different biological effectiveness of different sparsely ionising radiations is applicable at low doses and low-dose rates and is therefore relevant both to the definition of weighting factors for radiological protection purposes and to the use of X-rays in diagnostic medicine. It is not intuitive but the lower energy X-rays will be more biologically effective and, therefore, carry a somewhat higher risk than the more energetic X-rays.

6.6 CONCLUSIONS

The DNA double helix presents a well-defined three-dimensional target for the induction of a double strand break in the passage of a single ionising particle. Consequently, the quantitative appreciation of the biological effectiveness of different radiations makes use of the three-dimensional structure of the DNA double helix target and a theoretical description of the spatial energy deposition pattern of the radiation track at a nanometre resolution in three dimensions. The model permits a unifying analysis of several data sets measuring the variation of the linear (α) coefficient after exposure to different densely ionising radiations. These analyses reveal that the two energy depositions causing the double strand break are induced within a nanometre of the two DNA polynucleotide strands.

The model provides a mechanistic basis for the expectation that sparsely ionising radiations will not all have the same linear (α) coefficient or low-dose biological effectiveness.

7 Radiation-Induced Cancer

Cancer is assumed to arise from a somatic mutation and DNA double strand breaks are linked to cancer via mutations and aberrations but the exposed mutated cells have to survive to express the malignancy. Consequently, the equation for cancer induction has a component for mutation induction and a component for cell killing. The equation describes a curve which rises with linear–quadratic dose kinetics from the origin and passes through a peak to decrease at higher exposures when cell killing starts to dominate. The equation is applied to several data sets on the induction of cancer in small animals, demonstrating the influence of reduced dose rate and more densely ionising radiation on the level of cancer induction. The equation is applied to the induction of leukaemia in the atomic bomb survivors and is compared with the 'linear no-threshold' analysis, used in the regulation of radiological protection. The full equation fits the leukaemia data well but that fitting is completely dominated by the quadratic coefficient of the linear–quadratic dose–response. This means that it will be difficult to derive an accurate linear (α) coefficient which is needed to define the low dose, low-dose rate cancer induction relevant for radiological protection risk. The occurrence of cancer in children and young adults is discussed on the basis of a two-mutation model of cancer induction.

7.1 STOCHASTIC EFFECT

The health effects of radiation are normally divided into two categories – 'stochastic effects' and 'deterministic effects'. Stochastic effects, of which radiation-induced cancer is a prime example, are defined as effects where the probability of occurrence increases with radiation exposure dose but the severity of the effect is independent of the exposure dose. In other words, the cancer induced by a dose of 5 Gy is not more severe than the same cancer induced by a dose of 1 Gy.

Deterministic effects, on the other hand, of which lethality (Chapter 8) is a good example, do show an increasing severity with increasing dose: more small mammals are killed by a dose of 5 Gy than by a dose of 1 Gy.

The model is extended to investigate radiation-induced cancer in the following way. The link made from DNA double strand breaks to the three cellular effects – cell survival, chromosomal aberrations and mutations – is taken further to describe radiation-induced cancer, by considering which cellular effects might be involved. In the case of radiation-induced cancer, there are clear arguments to include both chromosomal aberrations and somatic mutations. Cell survival also plays a role because cancer results from the exposure of an organism, so that the affected cell has to survive to express the malignant phenotype. Figure 7.1 presents an expansion of Figure 3.1 showing the involvement of the three cellular effects in the induction of cancer.

There is abundant literature on the occurrence of somatic mutations in cancers (Stratton et al. 2009; Cancer Genome Atlas Research Network 2017a,b; Scarpa et al. 2017) and it is clear that chromosomal aberrations (Rowley 1973a,b; Grade

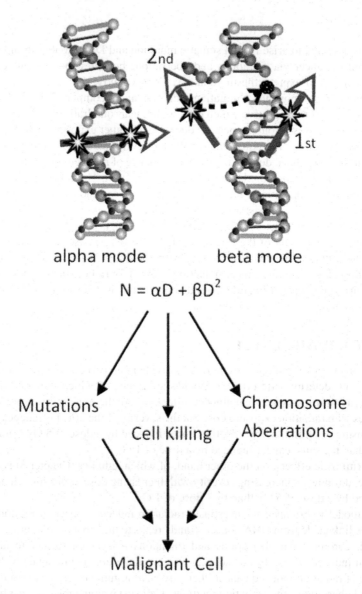

FIGURE 7.1 A schematic representation of the linkage chain from the molecular lesion, the DNA double strand break, to the three cellular radiobiological effects – cell inactivation, chromosome aberrations and mutation – indicating, that all three cellular effects are implicated in the induction of cancer.

et al. 2015) are involved in many of these mutations. It is also clear that chromosomal instability, possibly induced by radiation, is not the cause of the transition to malignancy but that it plays a role in driving metastasis (Bakhoum et al. 2018). It is, therefore, logical to invoke radiation-induced somatic mutations as a cause of radiation-induced cancer. However, because the cancer occurs in an irradiated organism, the malignantly mutated cell has to survive to express its malignant nature, and cell survival also has to be included in the equation, as it will affect the ability of the cell to express its malignancy.

Thus, the equation for mutation frequency per irradiated cell (M_I) as shown in Equation 2.5 is the equation which is relevant for cancer induction (CI) which becomes:

$$CI = \left(1 - \exp\left(-q_m\left(\alpha D + \beta D^2\right)\right)\right) \cdot \exp\left(-p\left(\alpha D + \beta D^2\right)\right) \tag{7.1}$$

where (q_m) is now the parameter relating DNA double strand breaks to the somatic mutation for the induction of malignancy.

Equation 7.1 describes a dose–response curve which increases from the origin with linear–quadratic dose kinetics but which flattens to a peak and decreases at higher doses as cell inactivation starts to dominate. This is shown in Figure 7.2.

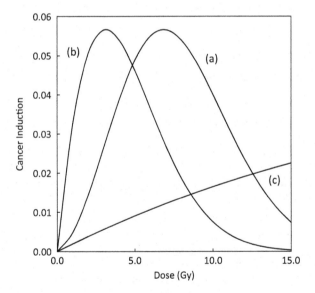

FIGURE 7.2 The graphical representation of Equation 7.1 for three different exposure situations. The curve (a) represents the cancer induction as a function of dose after an acute exposure of sparsely ionising radiation when the dose–effect has a balanced linear–quadratic form. The curve (b) represents the cancer induction as a function of dose after an acute or chronic exposure to densely ionising radiation when the dose–effect is dominated by the linear (α) coefficient. The curve (c) represents the cancer induction as a function of dose after a chronic exposure to sparsely ionising radiation when the quadratic (β) coefficient is zero.

Note that, as expected on the basis of the model, the peak height of the curves for acute sparsely and densely ionising radiation exposure is the same. This is a consequence of the role that radiation-induced DNA double strand breaks play in the creation of malignant mutations and also on the increasing cell inactivation, as has been discussed in Chapter 3, Section 3.5. This is the 'implied correlation' and, in this case, the peak height will depend on the relative values of the parameters (q_m) and (p) but be independent of the dose–effect coefficients (α) and (β).

Thus, by extending the model to describe the dose–effect relationship expected for radiation-induced cancer by modifying the equation for mutation frequency per irradiated cell, the model predicts that the Equation 7.1 should describe the dose–effect relationships for the induction of cancer and that the peak height of cancer induction (CI) should remain the same for different radiation treatments. The peak height will, of course, vary from one specific cancer to another and from one test system to another.

7.2   EXPERIMENTAL DATA

The following figures present some experimental data for the induction of cancer following exposure of animals to radiation.

In each of Figures 7.3 to 7.6, the peak height found for the sparsely ionising radiation (X or gamma rays) is the same as that found for the more densely ionising neutron irradiation even though the dose kinetics of the curves fitting the data are quite different. The peak height clearly depends on the type of cancer studied, the animal studied and the gender of the animal. The value of the peak height varies from 0.212 to 0.36.

Little importance should be given to the values of RBE_0 derived from these figures except to indicate the increasing importance of the linear (α) coefficient for the more densely ionising neutron irradiations. This is because, as discussed later, the fit of Equation 7.1 to the data, especially after exposure to sparsely ionising radiation, is dominated by the quadratic (β) coefficient.

Figures 7.3 to 7.6, which all show the same peak height for cancers induced by sparsely and densely ionising radiation, provide ample evidence that radiation-induced cancer arises from a somatic mutation (Chadwick and Leenhouts 2011a) in complete agreement with, and complementary to, the molecular biological studies of different cancers. Indeed, more recent cytological and molecular biological studies of the myeloid leukaemia induced in the CBA/H mice (see Figure 7.3) have revealed the role of deletions in chromosome 2 and a potential tumour suppressor gene on chromosome 2 (Bouffler et al. 1996; Silver et al. 1999; Cook et al. 2004) in the development of the radiation-induced leukaemia. Peng et al. (2009) have also found deletions on chromosome 2 in myeloid leukaemia in C57 BL/6 mice. It is fair to conclude that radiation-induced cytological damage in the form of DNA double strand breaks, leading to chromosomal aberrations or somatic mutations, is involved in the induction of cancer by radiation.

In fact, Equation 7.1 is really an approximation although it is sufficiently accurate for the analyses and the conclusions arising from them. In deriving the equation for cancer induction from the equation for the mutation frequency per irradiated cell, the number of cells in the organism at risk for the somatic mutation to malignancy has

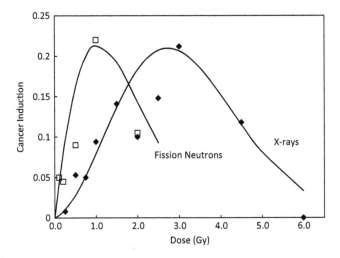

FIGURE 7.3 The induction of leukaemia in male CBA/H mice after exposure to X-rays (full diamonds) and fission neutrons (open squares) fitted with Equation 7.1 for cancer induction illustrating the same peak height at 0.212. The RBE_0, that is, the ratio of the linear coefficients, for the fission neutrons compared to the X-rays is 18 (data from Major and Mole 1978; Mole and Davids 1982; Mole et al. 1983).

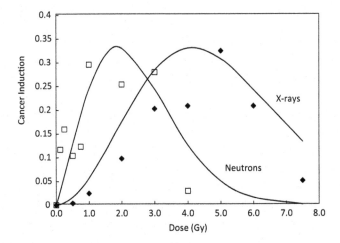

FIGURE 7.4 The induction of lung tumours in male mice after exposure to X-rays (full diamonds) and neutrons (open squares) fitted with Equation 7.1 for cancer induction illustrating the same peak height at 0.335. The RBE_0 for the neutrons compared with the X-rays is 28 (data from Coggle 1988).

been neglected. In addition, the chance that a malignantly mutated cell develops into a detectable cancer adds a further layer to the equation. This can be seen in the following example, starting with the equation for mutation frequency per irradiated cell:

$$M_I = \left(1 - \exp\left(-q\left(\alpha D + \beta D^2\right)\right)\right) \cdot \exp\left(-p\left(\alpha D + \beta D^2\right)\right) \qquad (7.2)$$

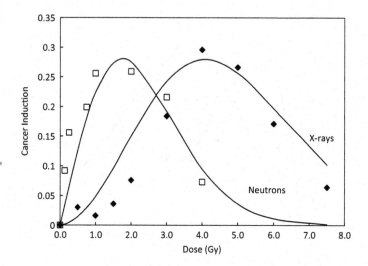

FIGURE 7.5 The induction of lung tumours in female mice after exposure to X-rays (full diamonds) and neutrons (open squares) fitted with Equation 7.1 for cancer induction illustrating the same peak height at 0.28. The RBE_0 for the neutrons compared with the X-rays is 40 (data from Coggle 1988).

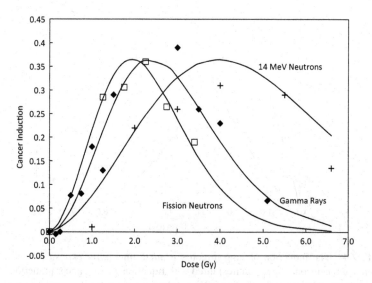

FIGURE 7.6 The induction of leukaemia in mice after exposure to gamma rays (full diamonds), fission neutrons (open squares) and 14 MeV neutrons (crosses) fitted with Equation 7.1 for cancer induction illustrating the same peak height at 0.36. The RBE_0 for fission neutrons compared with the gamma rays is 5.7 and the RBE_0 for 14 MeV neutrons is 4.0 (data from Upton et al. 1964).

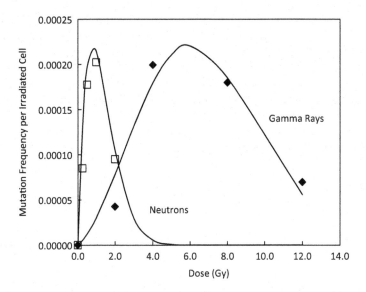

FIGURE 7.7 The induction of 8-azoguanine resistance mutations per irradiated cell in Chinese hamster cells by cobalt-60 gamma rays and fast neutrons plotted as a function of dose fitted using Equation 7.2 illustrating the same peak height (data from Richold and Holt 1974). These peaked curves were created from the original data by multiplying the mutation frequency per surviving cell by the cell survival. See also Chapter 3, Figure 3.19. Note the very small mutation frequency.

which has been fitted (see Figure 7.7) to some data from Richold and Holt (1974) who measured the induction of 8-azoguanine resistance mutations in Chinese hamster cells. They actually measured the mutations per surviving cell and also measured survival after gamma rays and fast neutron exposures. Using this data the mutation frequency per surviving cell has been converted to mutation frequency per irradiated cell, by multiplying the mutation frequency per surviving cell by the cell survival.

The value of (q) for the mutation in these cells is a factor of 0.0006 smaller than the value of (p) but if 1000 cells were at risk then the chance that one cell was mutated would be:

$$1000M_I = 1000\left(1 - \exp\left(-q\left(\alpha D + \beta D^2\right)\right)\right) \cdot \exp\left(-p\left(\alpha D + \beta D^2\right)\right). \quad (7.3)$$

All the vertical data are multiplied by 1000 and the peak height becomes 0.22 which is comparable with the cancer induction peaks. If the chance that one mutated cell develops into a cancer is taken into account, the equation becomes

$$CI = 1 - \exp\left(-m\left(\vartheta\left(1 - \exp\left(-q\left(\alpha D + \beta D^2\right)\right)\right) \cdot \exp\left(-p\left(\alpha D + \beta D^2\right)\right)\right)\right) \quad (7.4)$$

where (m) is the probability that a malignantly mutated cell grows into a cancer and (ϑ) is the number of cells at risk for the malignant mutation per organism.

This last modification of Equation 7.2 has no effect on the shape of the curves, unless the peak height is greater than 0.25. The modification merely means that the cancer induction cannot exceed 1.

The value of the parameter (q_m) used in the fitting of Equation 7.1 to the cancer induction data in Figures 7.3 to 7.6 includes the probability that a cell is mutated to a malignant phenotype. It also takes into account the number of cells, probably stem cells (Greten 2017; de Sousa e Melo et al. 2017), at risk for the mutation. It is this combination which makes the risk of radiation-induced cancer so much larger than the risk for a single mutation in an isolated cell.

7.3   DOSE RATE

Although the simplified Equation 7.1 for cancer induction predicts that, especially for sparsely ionising radiation and on the basis of the cellular effects, there should be a dose rate effect on the level of cancer induction, there does not appear to be much data to validate this. However, Mole and Major (1983) have measured the induction of leukaemia in CBA/H mice after protracted exposure to X-rays and this work reveals a clear effect of dose rate on cancer induction. In Figure 7.8, the cancer induction after exposure to acute (see Figure 7.3) and chronic X-rays is shown fitted to Equation 7.1 with the same linear (α) coefficients in both cases but with the quadratic (β) coefficients set to zero for the chronic exposure.

In order to illustrate the expected lack of a dose rate effect on cancer induction after densely ionising radiation, Figure 7.9 has been drawn using the data of Mole and his colleagues (Major and Mole 1978; Mole and Davids 1982; Mole et al. 1983; Mole and Major 1983) with the fitting of Equation 7.1 to the data. There are no data for the chronic fission neutron curve. This curve is derived from the fitting of Equation 7.1 to the acute neutron data but with the quadratic (β) coefficients set to zero.

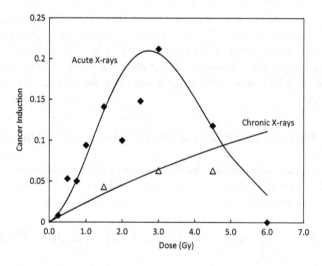

FIGURE 7.8   The induction of myeloid leukaemia in the CBA/H mouse after acute and chronic exposures to X-rays, analysed using Equation 7.1, demonstrating the reduced cancer incidence after chronic exposure, as expected from the model (data from Mole and Major 1983).

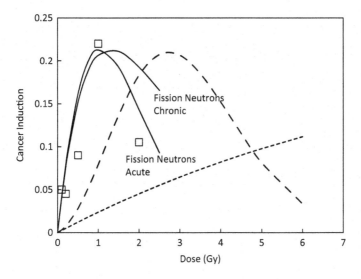

FIGURE 7.9 The expected negligible effect of reduced dose rate on the induction of myeloid leukaemia in CBA/H mice, derived from the fitting of Equation 7.1 to the acute neutron data, as shown in Figure 7.3 but, with the quadratic (β) coefficients set to zero. The dashed line and the dotted line are taken from Figure 7.8 for the acute and chronic exposure to X-rays and are included for comparison (data from Mole and Davids 1982).

These predictions of the effect of reduced dose rate on the induction of cancer by sparsely and densely radiation received support, while this book was being written, in a publication by Little (2018). This publication used a model independent analysis of the curvature of the induced effect and concluded from animal experiments (Grahn et al. 1992) that a reduction of effect on protracted exposure could be found in a variety of tumours in mice after sparsely ionising radiation exposure with dose reduction factors (DREF) of 1.2 to 2.3 (see Figure 7.8), but little or no dose rate reduction in effect could be found after densely ionising fission neutron radiation with a DREF of 0.0 to 0.2 (see Figure 7.9). Independent of the analysis made by Little, the work done by Grahn and his colleagues (Grahn et al. 1992) demonstrates a sparing effect of protracted sparsely ionising radiation exposures on cancer induction in a variety of tumours in B6CF$_1$ mice.

There are also some data from Ullrich et al. (1987) which show a reduced induction of lung adenocarcinoma and mammary adenocarcinoma in female BALB/c mice following gamma exposure. Although the studies only cover a restricted dose range up to 0.5 Gy, the results reveal a linear–quadratic curve for acute exposure and a linear curve for chronic exposure with the same linear (α) coefficients in both cases.

The sets of data from Mole and Major (1983), Grahn et al. (1992) and Ullrich et al. (1987) reveal a reduction of cancer incidence for different cancers going from an acute exposure to a chronic exposure of sparsely ionising radiation, in accordance with the expectations of the model. These experimental results are important for radiological protection as they establish the expected dose rate effect for the induction of cancer.

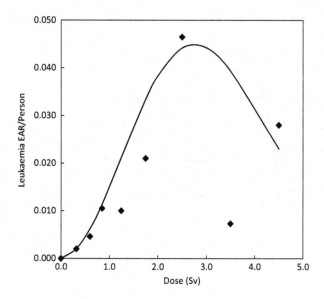

FIGURE 7.10 Equation 7.1 fitted to the EAR (excess absolute risk) per person for leukaemia in Japanese atomic bomb survivors (data from Pierce et al. 1996).

7.4 HUMAN DATA

The human data on the induction of cancer by ionising radiation come essentially from the Life Span Study of Japanese atomic bomb survivors. Figure 7.10 shows that Equation 7.1 can be successfully fitted to the data for leukaemia excess absolute risk (EAR) per person over the full weighted bone marrow dose range up to 4.5 Sv. The data are taken from an article by Pierce et al. (1996) because the more recent publications only cover the dose range up to 2 Sv (Ozasa et al. 2012), probably concentrating on the low-dose region to define the initial slope.

Many of the analyses of the data from the atomic bomb survivors made by regulatory bodies, including that of the International Commission on Radiological Protection (ICRP), have concentrated on only the rising part of the data, ignoring the reduced cancer induction at higher doses. The earlier analyses chose a linear fitting through the origin to the data, not unsurprisingly (see Figure 7.11), and this, together with an apparently linear dose response for the solid cancer data, gave rise to the 'linear no-threshold' (LNT) concept of radiation risk to cancer.

The ICRP, conscious of the dose rate effect in cellular radiobiology, applied a dose and dose rate effectiveness factor (DDREF) of 2 to derive the radiation risk for chronic sparsely ionising radiation. The combination of the LNT concept with a DDREF of 2 was used to derive a conservative approach to radiological protection and to provide a 'prudent basis for the practical purposes of radiological protection' (ICRP 2007).

Although there are no human dose rate data comparable to the animal data shown in Figure 7.8, Howe (1995) has made an epidemiological study of people exposed to many small doses of X-rays in lung fluoroscopy examinations and shown that under this fractionated exposure scheme there was apparently no additional risk of cancer

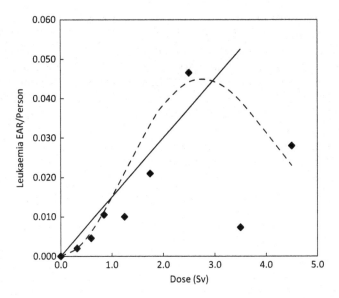

FIGURE 7.11   The data shown in Figure 7.10 with a linear fitting to the rising portion of the data.

at a total dose of 1 Sv. This study suggests, at least, a sparing effect of chronic exposure in human cancer.

In contradiction to the approach taken by ICRP and other regulatory bodies, the LNT concept has been challenged over the years by those who think that there is no induction of cancer at low doses and that radiation-induced cancer only occurs above a threshold dose (Becker 1997; Bond et al. 1996; Calabrese 2017; Tubiana 2000).

More recent analyses of the atomic bomb survivor data have investigated the indication of curvature evident in the low-dose region of the leukaemia data because, as Pierce et al. (1996) note, 'There is a statistically significant upward curvature in the ... dose range 0 – 3 Sv'. Rühm et al. (2016) have also looked at the low-dose curvature in all the different solid cancer data published by Ozasa et al. (2012) and found significant upward curvature in all the solid cancers.

With the benefit of a model developed from a molecular lesion, the DNA double strand break, to give an analytical equation for mutation frequency per irradiated cell that is extended to describe radiation-induced cancer arising from a somatic mutation with fitting of animal data and human data, a different understanding of the cancer induction process can be gained. On the basis of this understanding, several statements can be made:

- The dose–effect relationship for acute exposure to sparsely ionising radiation will be linear–quadratic from zero dose up but will saturate to a peak and decrease at higher doses as cell inactivation of the malignantly mutated cells begins to dominate.
- At low doses and low-dose rates, the dose–effect relationship will increase linearly from zero dose up and be defined by the linear (α) coefficient.

- This means that there will be a sparing effect of a lower dose rate on the cancer induction frequency (see Figure 7.8).
- It also means that there will not be a threshold dose below which exposure to ionising radiation will not carry some risk of cancer induction. This is because all ionising radiations are capable of inducing a DNA double strand break in the passage of a single ionising track, that is, in the alpha mode of double strand break induction quantified by the linear (α) coefficient.
- The dose–effect relationship for acute and chronic exposure to densely ionising radiation will be dominated by the linear (α) coefficient and will rise almost linearly from the origin up to a peak value before decreasing when cell inactivation begins to dominate.
- There will be very little effect of reduced dose rate on the cancer induction by densely ionising radiation.
- The peak height of cancer induction and to some extent the values of the linear (α) and quadratic (β) coefficients will vary from cancer type to cancer type and this means that data from different cancers should not be merged, for example, in the grouping 'all solid cancers', and should not be analysed using the Equation 7.1. Each cancer type should be analysed separately.
- The increasing part of the cancer induction data should not be analysed using a linear relationship through the origin, the so called 'linear no-threshold' concept. The rising portion of the cancer induction data will be linear–quadratic at low doses before saturating to the peak.

However, although Equation 7.1 should be the method of choice for the analysis of cancer induction data and good fits will be obtained over the complete dose range, unfortunately, the fitting for acute, sparsely ionising radiation exposures will be totally dominated by the quadratic (β) coefficient and the errors in the linear (α) coefficient will be large. This means that very little information about the risk for cancer induction from chronic exposure to sparsely ionising radiation will be gained from the fitting of Equation 7.1 to acute exposure data such as the atomic bomb survivor data. This is demonstrated in Figure 7.12 which shows the fitting to leukaemia data from the atomic bomb survivors made in Figure 7.10 with additional curves (dashed and dotted) where the linear (α) coefficients have been altered by plus and minus a factor of three. The fit looks good in all three cases because the quadratic (β) coefficient dominates (Leenhouts and Chadwick 2011). The straight lines represent the three different low-dose rate EAR dose–effect curves which are predicted when the quadratic (β) coefficients are set to zero for the dashed, solid and dotted curves, respectively.

In addition to this, the role of the potentially different weighting factors for different sparsely ionising radiations in radiological protection must be considered (see Chapter 6). The gamma rays from the atomic bomb were very energetic and would be expected to have a lower linear (α) coefficient compared with cobalt-60 and caesium-137 gamma rays and various X-rays from 250 kVp down to lower energies. It has been shown that the linear (α) coefficient of linear–quadratic model fitting of dose–effect data for unstable chromosome aberrations varied by a factor of 2–3 between gamma rays and X-rays in human lymphocytes exposed *in vitro*

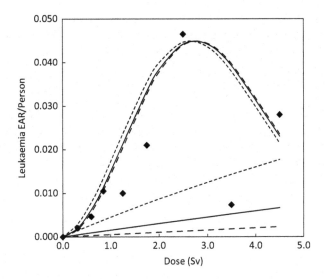

FIGURE 7.12 An illustration of the insensitivity of the fitting of Equation 7.1 to leukaemia EAR data from the atomic bomb survivor study. The solid peaked line is the fitting shown in Figure 7.10, the dashed line has a linear (α) coefficient three times smaller than that of the solid line, and the dotted curve has a linear (α) coefficient three times greater than the solid line. The lower three more linear curves show the predicted chronic exposure EARs when the quadratic (β) coefficient is set to zero in each of the peaked curves. These chronic curves show just how large an error in the linear (α) coefficient could be.

(Lloyd et al. 1986). Similarly, it has been suggested that the risk of mammography X-rays might be underestimated by a factor of 4 compared to gamma rays based on RBE values found in *in vitro* cell studies (Heyes et al. 2006; 2009; Heyes and Mill 2004; Frankenberg et al. 2002). In fact, careful cellular radiobiological studies using survival, chromosomal aberrations and somatic mutations to determine comparable values of the linear (α) coefficients, offer the optimum way of defining maximum RBE_0 values for a wide range of sparsely ionising radiations and they are badly needed for radiological protection guidelines.

The itemised points, together with the poor prognosis for a useful analysis of the atomic bomb survivor data to derive a low dose, low-dose rate risk shown in Figure 7.12, are all aimed at the various attempts to determine the chronic radiation risk from sparsely ionising radiation and, consequently, they are relevant for radiological protection.

In conclusion, on the basis of a somatic mutation leading to cancer, for which there is increasing evidence from molecular genetics as well as, independently, from the same peak height in the animal cancer data, the model provides an Equation 7.1 for cancer induction covering the full dose range. However, the lack of sensitivity of the analysis of atomic bomb survivor data to the value of the linear (α) coefficient means that it will be difficult to derive a value for the chronic radiation risk from the atomic bomb survivor data. As a result, the LNT fitting of the rising portion of the data has no relationship to the chronic radiation risk. Nevertheless, the model and the equation do predict that the chronic radiation risk of cancer increases from the origin linearly in proportion with the dose so that each small exposure brings a small, but positive, risk of cancer. There is no threshold dose below which a small dose does not carry some risk.

The actual chronic risk predicted by Equation 7.1 only applies to short-term chronic exposures when the cell population at risk can be assumed to remain constant. Exposures over many years, such as those accumulated by radiation workers, may be accompanied by changes in the cell population at risk and this can influence the value of the long-term chronic exposure risk. Epidemiological studies of chronically exposed populations supported by good dosimetry, similar to those of Leuraud et al. (2015) and Richardson et al. (2015), offer the best opportunity of determining the risk of chronic long-term exposure to ionising radiation.

7.5   CANCER IN CHILDREN AND YOUNG ADULTS

Cancer is a disease of old age with a peak incidence at around 75 years of age and yet there are cancers occurring in children and in young adults. The radiation model developed here can offer some insight into this phenomenon.

The important role of somatic mutations in stem cells in the development of cancer is no longer in doubt (Stratton et al. 2009; Cancer Genome Atlas Research Network 2017a,b; Scarpa et al. 2017) and the multi-stage nature of cancer has long been discussed ((Armitage and Doll 1954; Burch 1960). Recent studies of childhood cancers (Grobner et al. 2018; Ma et al. 2018; Bandopadhayay and Meyerson 2018) have indicated the role of germ-line mutations. It appears, therefore, that cancer arises as an accumulation of somatic mutations.

In his study of the incidence of retinoblastoma in children, Knudsen (1971, 1985, 1991) concluded that a two-mutation pathway was involved. This conclusion was later confirmed by cytological and mutagenic studies of the retinoblastomas (Knudsen et al. 1976; Yunis and Ramsay 1978; Sparkes et al. 1980; Gardner et al. 1982; Dunn et al. 1988; Stahl et al. 1994). On the basis of the retinoblastoma results, Moolgavkar and Knudsen (1981) developed a two-stage model of carcinogenesis which includes two mutation steps and also includes 'promotion' of intermediate cells on the road to a malignant cell that eventually develops into a tumour. This two-stage model of Moolgavkar and Knudsen has been modified here by using the cellular model of radiation action and the DNA double strand break as the critical mutational event, essentially to investigate its implication for radiation risk (see Chapter 9) (Chadwick et al. 1995, 2002; Leenhouts 1999; Leenhouts and Brugmans 2000; Leenhouts and Chadwick 1994a,b, 1997; Leenhouts et al. 1996, 2000). Figure 7.13 presents a schematic representation of the modified two-stage model.

The model covers the whole of the 'lifetime' and starts with a population of 'stem cells' which are targets for the 'initiation' mutation (μ_1) to create a single 'intermediate cell'. This single 'intermediate cell' undergoes division, a 'promotion' (ε) process which, over time creates an exponentially increasing population of 'intermediate cells'. These 'intermediate cells' are all targets for a second 'conversion' mutation (μ_2) which creates a single 'malignant cell'. The 'malignant cell' divides, also

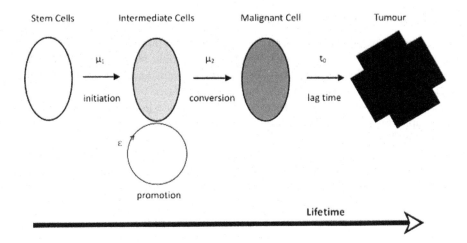

FIGURE 7.13 A schematic representation of the two-stage cancer model developed from the work of Moolgavkar and Knudsen (1981) where the two mutations (μ_1, μ_2) are assumed to arise, in the radiation version, from DNA double strand breaks.

exponentially, over 'lag time' (t_0) into a detectable tumour. (The reasoning remains unchanged even if there are more than two somatic mutations in the cancer process.)

The important thing to realise, at this point, is that *time* is deeply involved in many of the steps in cancer development. As a baby grows into an adult, time is involved in the development of a 'stem cell' population which is a target population for the 'initiation' mutation. The 'promotion' division of 'intermediate cells' to create an ever growing target population for the second 'conversion' mutation also takes time. Finally, the division of the 'malignant cell' during the 'lag time' creates a detectable tumour. These processes take place during the 'lifetime'.

The second thing to realise is that the 'initiation' mutation and the 'conversion' mutation both involve serious double strand damage to the DNA of the cells.

The third thing to realise is that it is not possible to know, *a priori*, whether the 'initiation' mutation can be also the 'conversion' mutation or not and also whether the 'conversion' mutation can be an 'initiation' mutation or not. The sequence in time of the mutations is not defined.

The fourth thing to realise is that there may be several different 'initiation' mutations and several different 'conversion' mutations which can ultimately lead to the same type of tumour.

7.5.1 CHILDHOOD CANCERS

If a 'germ-line mutation' is involved, as appears to be the case in many childhood cancers, all the cells in the body carry that mutation which means that they are all in the 'intermediate cell' stage. In other words, the germ-line mutation skips the first stage of the cancer process and the occurrence of cancer at a young age can be anticipated. This is probably just one reason for the occurrence of childhood cancer.

7.5.2 CANCER IN YOUNG ADULTS

Cancer in young adults is frequently associated with a family history of early cancers which suggests that a genetic predisposition could be involved. Although a 'germ-line' mutation might be associated with the occurrence of cancer at a young age, cancer in young adults can also be assumed to arise from a more rapid transit through the two-stage cancer process than that of those getting cancer in old age. This rapid transit through the cancer process can arise as a consequence of one or two different situations.

Anyone inheriting a DNA double strand lesion repair deficiency gene will accumulate the 'initiation' mutation and the 'conversion' mutation (and any other mutations that may be on the cancer pathway) more rapidly in time and, thus, cancer will develop at a younger age than in those who have normal DNA double strand lesion repair. The DNA double strand lesion repair, rather than the more specific DNA double strand break repair, is identified because many chemical mutagens can cause DNA double strand lesions (see Chapter 11). Some of these lesions might become DNA double strand breaks which will be repaired by the same process of micro-homology-enabled repair (MHERR) described for the repair of DNA double strand breaks in Chapter 4 and which plays such an important role in the production of chromosomal aberrations.

The important consequence of this is that many people presenting with cancers between the ages of 25 and 55 years are probably carrying a DNA repair deficient gene and these people will be extra sensitive to radiation in radiotherapy and especially double strand break-inducing chemicals in chemotherapy. It is essential that oncologists appreciate and are aware of this. It is interesting to note that the breast cancer genes, BRCA1 and BRCA2, often found associated with breast and ovarian cancers in young women (Brianese et al. 2018), are both considered to inhibit the repair of DNA double strand breaks and are implicated in the DNA recombination repair process (Cejka 2017; Zahao et al. 2017). Further evidence for the role of DNA repair deficiency in early onset cancer has been documented by Kobayashi et al. (2013) and Encinas et al. (2018).

The occurrence of breast and ovarian cancers in women carrying a BRCA mutation some 10 to 30 years before the occurrence of sporadic cancers of the breast and ovary in non-carriers has been well documented and is almost certainly a consequence of an inherited DNA repair deficiency in BRCA carriers. Cancer therapy, with radiation or chemicals causing DNA double strand breaks following a standard therapy regime, will have a more aggressive effect in any patient with a DNA repair defect, including BRCA carriers. More cell killing and more chromosomal damage will occur in DNA repair deficient patients than in patients with a normal DNA repair efficiency, not only in the malignant cells but also in any healthy tissue cells affected by the treatment. Consequently, a first check on any young adult presenting with any cancer should be a test for a DNA repair deficiency or radiation and chemical mutagen sensitivity. An appropriately modified therapy treatment regime should then be selected.

It has recently (2019) been suggested by Prof. G. Evans of Manchester University that women with a family history of early breast cancers would benefit from routine mammography from the age of 30 instead of the standard age of 40. These extra mammography examinations could lead to early detection and improved treatment of malignancy. This suggestion has considerable merit but, in view of the probability that many of these people could carry a DNA damage repair deficiency gene, for example a BRCA gene, caution is needed in the use of ionising radiation for the extra examinations. Each mammography examination for these women leads to a small increase in radiation exposure to the target tissue, the breast, with a consequent, small increase in the risk for breast cancer development. Health authorities would be well advised, in these cases, to find a method for breast examination which does not use ionising radiation.

A more rapid transit through the cancer process might also arise in people living in a more mutagenic environment so that the concentration of mutagens experienced by the 'stem cells' and the 'intermediate cells' of the model will be higher and the 'initiation' and 'conversion' mutations will occur at a younger age than for people living in a less mutagenic environment. This might depend on lifestyle, for example, smoking, or it may depend on where you live, as in the case of radon concentration in houses built on granite.

Finally, any inflammatory process causing an increased 'promotion' turnover of the 'intermediate cells' could create a larger target population of these cells for the 'conversion' mutation and might lead to the earlier occurrence of malignancy. However, it is unlikely that this process would be associated with a DNA repair deficiency so that these young cancer patients would not normally be more sensitive to therapy regimes than older patients.

7.6 CONCLUSIONS

The model gives an interpretation of radiation-induced cancer by applying the equation for mutation frequency per irradiated cell to the concept of cancer arising from a somatic mutation. The equation applies to the full dose range and is fitted to several data sets on cancer induced in animals and man. However, in all cases of sparsely ionising radiation exposure, the fitting of the equation is totally dominated by the quadratic (β) coefficient and information about low-dose radiation effectiveness, therefore, has large errors.

The combination of the radiation model with the multi-step nature of cancer leads to the suggestion that young adults presenting with cancer could carry a DNA repair deficiency gene and be extra sensitive to standard radiotherapy regimes.

8 Radiation-Induced Lethality

DNA double strand breaks are linked to the lethality of an organism via cell killing. The dose-response curve for lethality of an exposed organism demonstrates a long threshold followed by a sharp decrease at higher doses. Lethality is ascribed to multi-cell inactivation of stem cells. The relationship between the equation for single cell killing, derived in Chapter 2, and lethality is presented. The effects of reduced dose rate of exposure and more densely ionising radiations previously seen in single cell survival are shown to be reflected in the analysis of mouse survival data. It is important to recognise that, although lethality data exhibit a threshold of no apparent effect over a considerable dose range, the associated single cell survival curves reveal that substantial levels of stem cell inactivation and, therefore, chromosomal damage and mutations, occur as the threshold extends to higher doses.

8.1 RADIATION-INDUCED LETHALITY

In the previous chapter, radiation-induced cancer was classified as a 'stochastic effect' defined as an effect of radiation, where the severity of the cancer was not influenced by the level of the exposure dose. In this chapter, radiation-induced lethality is classified as a 'deterministic effect' defined as an effect where its severity does increase with increasing exposure dose. The radiation sensitivity of a uniform population of animals is often identified in terms of the dose of radiation which kills 50% of the population within a given time frame. Thus $LD_{50}/30$ refers to the dose needed to kill 50% of a population within 30 days.

As before, the model uses the link from DNA double strand breaks to the three cellular effects – cell survival, chromosomal aberrations and mutations – to describe radiation-induced lethality by considering which cellular effects might be involved. An acute exposure of an organ or organism to substantial doses of radiation leads to observable deleterious biological effects within a relatively short period of time, such as, from a few days to some weeks. These deleterious effects arise as a result of gross damage to the exposed organ or organism which is caused by significant cell inactivation. Thus, in the case of radiation-induced organism lethality, or organ malfunction, the cellular effect which is clearly implicated is cell inactivation. Figure 8.1 presents an expansion of Figure 3.1 showing the involvement of cell inactivation in the induction of lethality.

The acute radiation exposure of an organism, such as a small mammal or the organ of an animal, to significant doses of penetrating ionising radiation leads to either death of the mammal or a malfunction of the organ as a result of multi-cell

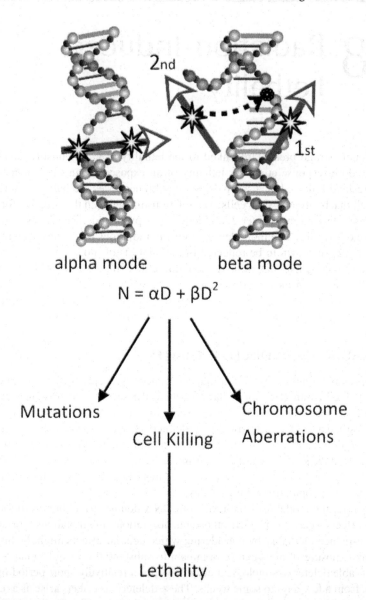

alpha mode   beta mode

$$N = \alpha D + \beta D^2$$

Mutations

Cell Killing   Chromosome
                      Aberrations

Lethality

FIGURE 8.1 A schematic representation of the linkage chain from the molecular lesion, the DNA double strand break, to the three cellular radiobiological effects – cell inactivation, chromosomal aberrations and mutation – indicating that only cell inactivation is implicated in radiation-induced lethality.

inactivation. If the amount of damage, which is dependent on the magnitude of the dose, does not cause the death of the organism or complete malfunction of the organ, cell renewal over time may lead to recovery and a continued functioning of the organism or organ. Consequently, these effects are not easily studied. However, in the case of a total body exposure of, for example, a small mammal, when death

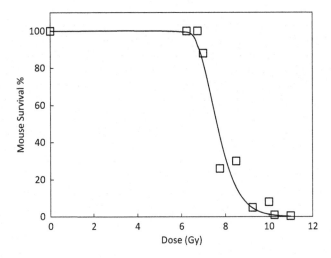

FIGURE 8.2 The effect of acute exposure to gamma radiation on mouse lethality, illustrating the long threshold from 0 to 6 Gy, where there is apparently no effect, followed by a precipitous decrease in mouse survival from 6 Gy to 10 Gy (data from Traynor and Still 1968). This sort of curve is typical for deterministic effects.

may be the resultant effect, studies have shown that the dose–effect relationship for these deterministic effects exhibits a long threshold followed by a steep decrease in survival. In the threshold region, no effect is apparent but, in the steep decrease, the severity of the observed effect, that is, the number of small mammals dying, increases with the increasing dose of radiation. Figure 8.2 presents a typical example of the dose–effect relationship for a deterministic effect, in this case mouse lethality, following a total body exposure to acute gamma rays (Traynor and Still 1968).

In extending the cellular effects model to provide an analytical approach to deterministic effects (Leenhouts and Chadwick 1983, 1989), it is assumed that the majority, if not all, of these effects arise as a consequence of multi-cell inactivation. This occurs when a substantial proportion of cells in an organ die with the result that the surviving stem cells for that organ are unable to repopulate it in time to prevent its malfunction and/or the death of the organism. Using this assumption, an equation can be derived for organ survival or organism lethality (L) from the equation for cell survival (S):

$$S = \exp\left(-p\left(\alpha D + \beta D^2\right)\right) \tag{8.1}$$

$$L = 1 - \left(1 - \exp\left(-p\left(\alpha D + \beta D^2\right)\right)\right)^{\nu} \tag{8.2}$$

where (ν) is related to the proportion of the stem cell population needed for organ repopulation such that, when an average proportion of less than (1/ν) of the original stem cell population survives, organ function is impaired. For example, if ν = 200, the organ fails when less than 1 in 200 or 0.5% of the stem cells survive.

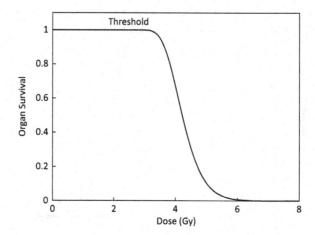

FIGURE 8.3 Equation 8.2 based on multi-cell inactivation, drawn for comparison with Figure 8.2 showing mouse survival after exposure to gamma radiation. Note the linear scale of the vertical 'Organ Survival' (lethality) axis.

It is important to realise that, although the acute radiation exposure renders the normal differentiated and no-longer dividing organ cells unable to function, it is the killing of the stem cell population for that organ which means that, once damaged, it cannot be repopulated and recover its normal function and this results in, potentially, the death of the organism.

Figure 8.3 presents lethality, that is organ or organism survival, according to Equation 8.2. Note the long threshold where there is no apparent effect followed by the steep decrease in survival to zero survival.

Figure 8.4 is drawn to illustrate the relationship between Equation 8.1 for single cell survival and Equation 8.2 for organ survival.

An important consequence of the lethality or organ survival being 'driven' by the single cell survival is that, although there is apparently no effect on the deterministic effect of lethality or organ survival on the long threshold dose, there is, in fact, a considerable amount of cell inactivation going on. It is also important to bear in mind that single cell inactivation is implicitly paired with increasing chromosomal damage and increasing mutations. The long threshold with no apparent effect on lethality or organ survival does conceal considerable tissue and organ damage.

The single cell survival curve, Equation 8.1, 'drives' the lethality or organ survival curve, Equation 8.2. This means that certain predictions can be made about dose rate effects and radiation quality effects on lethality or organ survival from the experience acquired on the effects that dose rate and radiation quality have on single cell survival.

8.2 THE EFFECT OF DOSE RATE

In Chapter 5, it was shown that the effect of reducing dose rate on Equation 5.1 for the induction of DNA double strand breaks is reflected in a decrease in the quadratic (β) coefficient. The explanation for the dose rate effect at the molecular level is that

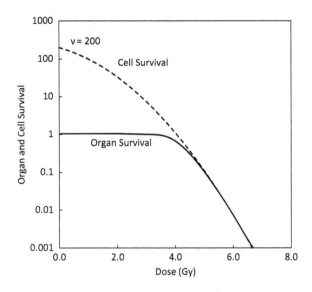

FIGURE 8.4 An illustration of the relationship between Equation 8.1 for cell survival and Equation 8.2 for lethality or organ survival. The cell survival 'drives' the lethality or organ survival which shows only the long threshold until the proportion of cells surviving reaches the level of 0.5% in this case. At this stage the surviving stem cells can no longer save the organ function and the lethality or organ survival starts to decrease following the curve of the stem cell survival. Note that both curves are drawn with a logarithmic scale on the 'Organ and Cell Survival' vertical axis.

the perfect repair of 'first' DNA single strand breaks occurs before a 'second' event converts them to DNA double strand breaks. This change in the quadratic (β) coefficient affects the single cell survival Equation 8.1 and carries through into the dose–effect Equation 8.2 for lethality or organ survival. In the protracted region of dose rate, the quadratic (β) coefficient is reduced and the effect of this can be reproduced in Equation 8.2 to show the expected effect of reduced dose rate on lethality or organ survival. Figure 8.5 illustrates this in a purely lethality graph and Figure 8.6 shows how the single cell survival 'drives' lethality.

Traynor and Still (1968) have investigated the effect of decreasing the dose rate of cobalt-60 gamma rays on mouse survival. The data, which are presented in Figure 8.7, illustrate the extension of the threshold as the dose rate decreases that is shown in Figure 8.5.

All the data have been combined into a single analysis using Equation 8.3 where the ($\beta(t)$) coefficient is assumed to vary with time of exposure (t) according to Equation 8.4. This relationship was previously derived in Chapter 5 on dose rate effects.

$$L = 1 - \left(1 - \exp\left(-p\left(\alpha D + \beta(t)D^2\right)\right)\right)^v \tag{8.3}$$

$$\beta(t) = \left(e^{-\lambda t} - 1 + \lambda t\right) \cdot 2\beta_\infty \big/ (\lambda t)^2 \tag{8.4}$$

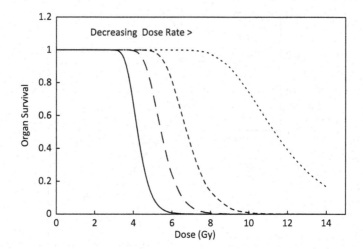

FIGURE 8.5 The effect of decreasing dose rate on lethality or organ survival according to Equation 8.2. The threshold is extended as the dose rate decreases and the value of the quadratic (β) coefficient is decreased. The dotted line is for a chronic exposure when the quadratic coefficient is zero.

The analysis reveals values of $p\alpha = 0.589$ Gy^{-1}, $p\beta_\infty = 0.0623$ Gy^{-2}, $\nu = 2300$ and the single strand break repair rate $\lambda = 0.025$ min^{-1} which gives a DNA single strand repair half-life of 28 minutes, comparable with 40 minutes measured directly in cells by Dugle and Gillespie (1975). Although these values of ($p\alpha$) and ($p\beta_\infty$) are compatible with values found for single cell survival, no significance can be attached to them

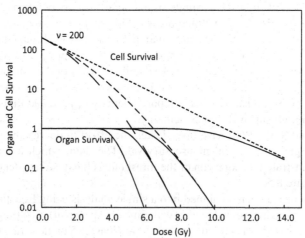

FIGURE 8.6 An illustration of how the single cell survival Equation 8.1 drives the lethality or organ survival curve, defined by Equation 8.2, as the dose rate is decreased. The single cell survival curves show the effect of the decreasing quadratic (β) coefficient which is reflected in the extension of the threshold to higher doses for the lethality or organ survival curves. The dotted curve defines single cell survival following a chronic exposure when the quadratic coefficient is zero. Note the logarithmic scale on the vertical axis.

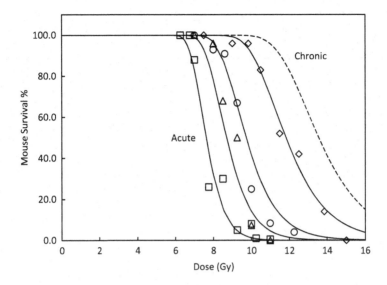

FIGURE 8.7 Mouse survival data ($LD_{50}/30$) (Traynor and Still 1968) measured at different dose rates from acute (120 Gy/h) (open squares), 8.4 Gy/h (open triangles), 4.02 Gy/h (open circles) to 1.25 Gy/h (open diamonds). The curves are drawn using the values derived from the analysis of the data with Equations 8.3 and 8.4. The dashed line illustrates the predicted mouse survival for a chronic exposure when ($\beta(t)$)$=0$.

because so little is known about the potential values of (ν) and ($1/\nu$), the proportion of stem cells needed for recovery of function.

8.3 THE EFFECT OF RADIATION QUALITY

In Chapter 6 it was shown that changes in the quality of radiation from, for example, sparsely ionising radiation, such as X-rays, to more densely ionising radiation, such as fast neutrons, leads to an increase in the linear (α) coefficient of the Equation 6.1 for the induction of DNA double strand breaks. This increase in the linear (α) coefficient affects the single cell survival Equation 8.1 and carries through into the Equation 8.2 for lethality or organ survival. In Figure 8.8 the effect of an increase in the linear (α) coefficient on the lethality or organ survival curve is illustrated in comparison with the acute curve for sparsely ionising radiation from Figure 8.3 and the chronic exposure curve for sparsely ionising radiation from Figure 8.5.

In Figure 8.9 the single cell survival curves for the sparsely ionising radiation acute and chronic exposures and the densely ionising radiation are drawn according to Equation 8.1 together with the lethality or organ survival curves drawn according to Equation 8.2.

Figure 8.10 presents data on mouse survival ($LD_{50}/5$) following acute X-ray exposure and acute fast neutron exposure (Hornsey 1973), analysed using Equation 8.2, where only the value of the linear ($p\alpha$) coefficient is allowed to vary.

The analysis of the data gives values for the X-ray exposure of $p\alpha = 0.64$ Gy^{-1}, and for the fast neutrons of $p\alpha = 2.18$ Gy^{-1}, with $p\beta = 0.0128$ Gy^{-2}, and $\nu = 150{,}000$. Again,

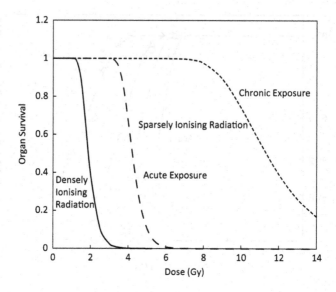

FIGURE 8.8 The effect of densely ionising radiation on the lethality or organ survival curve (solid line) compared with the acute lethality curve for sparsely ionising radiation from Figure 8.3 (dashed line) and the chronic lethality curve for sparsely ionising radiation from Figure 8.5 (dotted line). The lethality curve following densely ionising radiation exposure exhibits a shorter threshold.

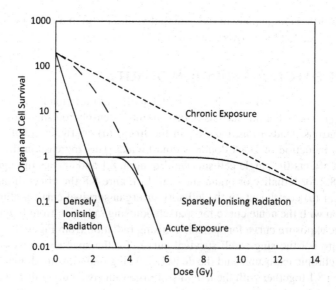

FIGURE 8.9 An illustration of how the single cell survival Equation 8.1 drives the lethality or organ survival Equation 8.2 as the radiation quality changes from sparsely to densely ionising radiation (solid line). The single cell survival curves show the effect of the increasing linear (α) coefficient which is reflected in the shortening of the threshold for the lethality or organ survival curves to lower doses. The dotted curve defines single cell survival following a chronic exposure to sparsely ionising radiation when the quadratic (β) coefficient is zero. Note the logarithmic scale on the vertical axis.

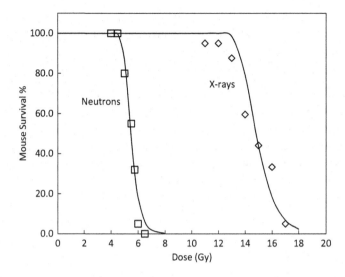

FIGURE 8.10 Mouse survival data (Hornsey 1973), measured after acute X-ray (open diamonds) and acute fast neutron exposures (open squares). The curves are drawn using values derived from the analysis using Equation 8.2. The data illustrate the considerably shorter threshold following the fast neutron exposure.

although these values for (pα) and (pβ) are compatible with those found for single cell survival, no significance can be attached to them because so little is known about the potential values of (v) and (1/v), the proportion of stem cells needed for recovery of function. The value of (v) for mouse survival here is some 65 times larger than that found in the analysis of the dose rate effect.

8.4 CONCLUSIONS

The extension of the model from DNA double strand breaks through cellular effects, in this case cell inactivation, leads to the derivation of an Equation 8.2 to describe lethality or organ malfunction based on the radiation-induced killing of a critical proportion of the stem cells needed to provide recovery of organ or organism function. The Equation 8.2 is clearly driven by the equation for single cell inactivation (Equation 8.1) and the influence of dose rate and radiation quality on single cellular effects, previously considered in Chapters 5 and 6, can be found reflected in the influence of dose rate and radiation quality on lethality or organ malfunction.

The effect of decreasing dose rate on single cell survival is explained mathematically by the decreasing value of the quadratic (β) coefficient in Equation 8.1, as a result of the perfect repair of DNA single strand breaks during the longer exposures. The result of this, when transposed to Equation 8.2 for lethality, is to lengthen the threshold of 'apparent' no effect and reduce the steepness of the reduction in organism or organ survival at the end of the threshold.

The effect of densely ionising radiation exposure on single cell survival, compared with sparsely ionising radiation, is explained mathematically by an increase in the linear (α) coefficient in Equation 8.1. The result of this when transposed to

Equation 8.2 for lethality is to considerably shorten the threshold of no 'apparent' effect and to steepen the reduction in organism or organ survival at the end of the threshold.

The effects of applying other exposure conditions, such as increased oxygen supply, which alter the sensitivity of the cells to the radiation and, consequently, alter the values of the coefficients (α) and (β) in the single cell survival Equation 8.1, will also be reflected in the concomitant modification of lethality, given by Equation 8.2.

The model provides a comprehensive understanding of the effects of radiation on the deterministic effects such as lethality or organ malfunction.

However, it remains difficult to make a unique evaluation of a set of lethality or organ malfunction data using Equation 8.2 because of the lack of knowledge about the potential value of ($1/v$), the proportion of stem cells needed to achieve recovery of function. In a uniform population of animals, for example, the value of (v) can be expected to be more or less constant but it could be dependent on gender, age, fitness, species, and so on. Care is therefore needed in making an analysis of lethality or organ malfunction data using Equation 8.2 and little significance should be given to the parameter values derived from the analysis.

It is important to recognise that, although lethality or organ malfunction data and Equation 8.2 exhibit a threshold of no 'apparent' effect over a considerable dose range, the associated single cell survival curves reveal that substantial levels of stem cell inactivation and, therefore, chromosomal damage and mutations which increase as the threshold extends to higher doses, will be affecting the stem cells.

9 Radiological Protection

The current risk of low-dose radiation exposure is derived from cancer incidence data in the atomic bomb survivors using the concepts of 'linear no-threshold' analysis of the increasing cancer incidence data and the application of a 'dose and dose rate factor' (DDREF) of two to take account of the reduced effectiveness of low-dose rate on cancer induction. The equation for cancer induction, derived in Chapter 7 and fitted to the atomic bomb survivor data, shows a rising, sloping 'S' shape close to the 'linear no-threshold' straight line but the fitting of the cancer induction equation is completely dominated by the quadratic coefficient of the dose response. This means that the atomic bomb survivor data give no information about the low-dose radiation effectiveness. Consequently, the 'linear no-threshold' concept also provides no information about the low-dose effectiveness and is redundant and the DDREF, which acts in the 'linear no-threshold' analysis, has no function. The low-dose radiological risk needs urgent revision. A way forward is suggested taking advantage of recent epidemiological studies and using a two-mutation model for cancer induction. The model predicts that the spontaneous cancer incidence will affect the low-dose risk and makes it necessary to distinguish between a short-term exposure and a long-term exposure. Hereditary effects are briefly mentioned.

9.1 INTRODUCTION

Recommendations from the International Commission on Radiological Protection (ICRP) (ICRP 2007, 1991) provide the basis for safe radiological protection practices worldwide. The current radiation risk estimation on which the recommendations are based relies on three concepts: the linear no-threshold (LNT) concept for the analysis of cancer induction in the atomic bomb survivors; a dose and dose rate effectiveness factor (DDREF) of 2 to apply the risk to chronic exposure regimes; and a weighting factor to convert dose in Gray (Gy) to dose equivalent in Sievert (Sv).

The model for radiation effects arising from the induction of DNA double strand breaks, presented in the previous chapters, throws each of these three concepts into doubt but presents an alternative solution for the derivation of low-dose and low-dose rate radiation risk.

9.2 LINEAR NO-THRESHOLD CONCEPT

The linear no-threshold concept assumes, conservatively, that the sparsely ionising radiation of the atomic bomb energetic gamma rays induces cancer in a single-hit process from zero dose up. The current radiation risk recommended by the ICRP and other regulatory bodies is based on a linear no-threshold analysis of the induction of

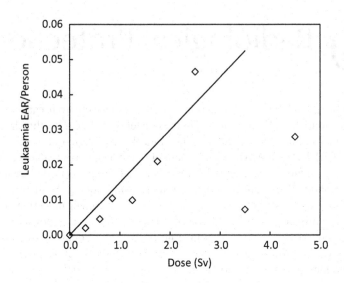

FIGURE 9.1 A plot of the excess absolute risk (EAR) per person for leukaemia in the atomic bomb survivors fitted with a straight line using the linear no-threshold concept (data from Pierce et al. 1996).

leukaemia in the atomic bomb survivors similar to the one which is shown schematically in Figure 9.1.

This linear fitting of the data by ICRP is not surprising in view of the increasing leukaemia data up to 3 Sv but it ignores the decreasing data above 3 Sv. When the radiological risk estimation was initially made, the regulatory authorities did not have a mechanism for the induction of cancer by radiation, and the derivation of risk preceded the recent research showing the involvement of somatic mutations in the origin of cancer.

With the insight offered by a mechanistic cause, that is, a somatic mutation, for the induction of cancer and a model for radiation action based on the induction of DNA double strand breaks giving an equation for the induction of somatic mutations per irradiated cell, the leukaemia data can be analysed over the complete dose range.

The approximate Equation 7.1 for the induction of cancer derived in Chapter 7 is

$$CI = \left(1 - \exp\left(-q\left(\alpha D + \beta D^2\right)\right)\right) \cdot \exp\left(-p\left(\alpha D + \beta D^2\right)\right). \tag{9.1}$$

The fitting of this equation to the leukaemia data is shown in Figure 9.2.

Although the equation gives a reasonably good fit to the data, the problem with the fitting of this type of equation is that the rising portion of the curve is dominated by the quadratic (β) coefficient and the fitting leaves a very large error in the value of the linear (α) coefficient which is relevant for the low-dose and low-dose rate cancer induction. This was illustrated in Figure 7.12. In Figure 9.3, this is reproduced up to a dose of 3.5 Sv together with the linear no-threshold straight line but the three curves drawn with the quadratic (β) coefficient equal to zero in Figure 7.12 have been omitted. The fact that the linear (α) coefficients of the three linear–quadratic

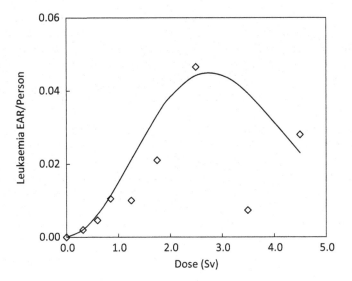

FIGURE 9.2 The fitting of Equation 9.1 to the excess absolute risk for leukaemia per person (data from Pierce et al. 1996).

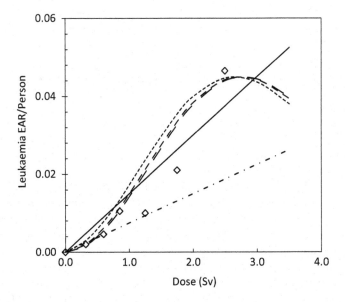

FIGURE 9.3 A comparison of the linear no-threshold fit (continuous straight line) to the leukaemia EAR/person from the atomic bomb survivors (data from Pierce et al. 1996), with three fits using Equation 9.1 in which the linear (α) coefficient is varied by a factor of 9. The linear no-threshold line divided by the DDREF of 2 is also shown (dash double-dotted line).

curves shown in Figure 9.3 vary by a factor of 9, leads to the conclusion that not one of the fittings including the LNT straight line, even with a DDREF of 2, provides any information about the low-dose and low-dose rate effectiveness of the atomic bomb radiation. In other words, there is no information from any of these types of analyses that is useful for the derivation of low-dose radiation risk.

The use of the linear no-threshold concept for the derivation of radiation risk should, therefore, be abandoned.

This does not mean that the low-dose, low-dose rate radiation risk is not linear. It is, but it is defined by the linear (α) coefficient of Equation 9.1 which implies that the low-dose, low-dose rate radiation risk will increase in proportion with dose from zero dose up. However, the linear (α) coefficient derived from the fitting of Equation 9.1 to cancer data following an acute exposure has been shown to have large errors and will not be useful for application in radiological protection. In this respect, it has to be remembered that the analysis of a variety of atomic bomb-induced cancers by Rühm et al. (2016) in the dose range up to 2 Sv found consistent curvature of the dose–effect responses but the linear–quadratic fittings to the data had large errors in the linear coefficient.

The conservative philosophy, adopted by the ICRP and other regulatory bodies, that each increment of radiation dose, no matter how small, brings with it a consequent increment in the risk of developing a cancer is, however, confirmed by the model.

The acceptance that the induction of DNA double strand breaks is the crucial mechanism by which radiation induces biological effects, implies that low doses of radiation will cause a linear increase in biological effect from zero dose up via the 'alpha' mode of radiation action.

9.3   THE DOSE AND DOSE RATE EFFECTIVENESS FACTOR

When the current radiological risk estimation was made, the ICRP and other regulatory bodies were aware that cellular radiobiological studies were revealing a sparing effect of low-dose rate exposure. Therefore, they applied a DDREF of 2 to derive a risk for low-dose and low-dose rate sparsely ionising radiation exposures. Starting from a 'linear no-threshold' straight line analysis of the acute atomic bomb survivor data, a constant value for the DDREF provided a linear dose–effect relationship for the risk from low-dose and low-dose rate sparsely ionising radiation exposures. Even though there was no mechanistic concept behind either the acute linear no-threshold analysis or the reduction to the low dose, chronic exposure regime, the linear dose–effect relationship for the chronic exposures provided a useful basis for the further development of radiological protection philosophy.

However, with the advent of models of radiobiological effects which promulgated a linear–quadratic approach to the analysis of cellular data (Kellerer and Rossi 1972; 1978; Chadwick and Leenhouts 1973a, 1981) questions were raised about the constant value of the DDREF and it was shown that the DDREF would, in fact, vary considerably with dose (Rossi 1990; Chadwick 2017). In addition, more recent studies of the atomic bomb survivor data revealed a clear quadratic curvature in many of the different cancers studied in the dose range up to 2 Sv and this also raised the question of the value of the DDREF (Rühm et al. 2016).

In fact, without the LNT concept, there is no need for a DDREF and the concept should be abandoned.

9.4 THE WEIGHTING FACTOR OF SPARSELY IONISING RADIATION AND DOSE EQUIVALENT

The ICRP and other regulatory bodies apply a radiation weighting factor (w_R) of 1 for all sparsely ionising radiations to convert the dose measured in Gray (Gy) to the dose equivalent in Sievert (Sv). By using weighting factors greater than 1 for densely ionising radiations, the exposures arising from different radiations can all be summed up in Sieverts so that a uniform risk, estimated in Sieverts, could be applied to all radiations. The weighting factor for sparsely ionising radiations is important because it provides the basis for the estimation of the dose equivalent of an exposure and the derivation of radiological risk from that exposure.

Several articles which cast doubt on the single value of the radiation weighting factor for sparsely ionising radiations have already been mentioned and discussed (Lloyd et al. 1986; Frankenberg et al. 2002; Heyes and Mill 2004; Heyes et al. 2006, 2009) with respect to gamma rays and X-rays, and Bridges (2008), Fairlie (2007), Goodhead (2009) and Harrison (2009) have all raised the issue of low-dose effectiveness for beta-emitting radioisotopes, especially tritium. Variations of up to the value of 4 have been implicated for gamma rays and X-rays and values of between 2 and 3 for beta emitters.

ICRP (2003) have also reviewed RBE and the radiation weighting factor and decided to maintain the weighting factor at 1 for different sparsely ionising radiations. Recently, the US NCRP (2018) reported on the effectiveness of low energy photons and electrons with respect to cancer induction. Their report introduces an effectiveness ratio for five different low energy groups suggesting a value increasing from 1 to 1.5 for tritium and 2.5 for photons of less than 1.5 keV. The report does not, however, make any recommendations for a change to the value of the radiation weighting factor for photons and electrons currently used in radiation protection regulation.

The model proposed for the induction of double strand breaks by ionising radiation in the alpha and beta modes of action (see Figure 9.4) provides a reason for the differences in the low-dose effectiveness of different sparsely ionising radiations.

In Chapter 6, it was shown that the choice of the DNA double strand break as the crucial radiation-induced lesion presents the DNA double helix molecule as an accurately defined, three-dimensional target for radiation track action. Analyses demonstrated that the choice of this molecular lesion meant that all forms of ionising radiation produced individual tracks with sufficient energy to break both DNA helix strands (the alpha mode of breakage, see Chapter 1) giving rise to the linear (α) coefficient component of radiation effect at the cellular level. Densely ionising radiations were seen to be very efficient inducers of the DNA double strand break in a single track and registered large linear (α) coefficient values in cellular effects with very little influence from the quadratic (β) coefficient. Sparsely ionising radiations, which essentially generate energetic electron scattering in their passage through tissue, were seen to be relatively poor inducers of the DNA double strand break in a

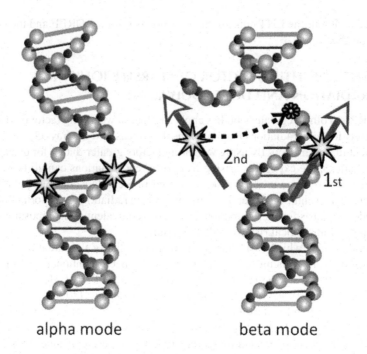

FIGURE 9.4 A schematic representation of the two modes of radiation action breaking the DNA helix.

single track and registered dose–effect curves with smaller linear (α) coefficients but a strong influence of the quadratic (β) coefficient.

The induction of a DNA double strand break in the passage of a single electron track (the alpha mode of breakage) requires two energy deposition events in some 1.2 nm along the track, the distance between the two DNA strands. Energetic scattered electrons depositing ionisation events at distances of more than 2–3 nm along a tortuous path will be very poor inducers of a DNA double strand break in the alpha mode. However, the work with ultrasoft X-rays (Cox, Thacker and Goodhead 1977; Goodhead, Thacker and Cox 1979) demonstrated the relatively high efficiency, with which the low energy electron tracks, induced by the ultrasoft X-rays, caused cellular effects with large linear (α) coefficients comparable to more densely ionising radiation. This indicates that the low energy scattered electrons near the ends of sparsely ionising radiation tracks will be efficient in breaking the DNA helix in the alpha mode. The relative effectiveness of different sparsely ionising radiations depends, therefore, on the proportion of radiation dose which is contributed by the effective low energy electron track ends. The larger that proportion of dose is, the larger the low-dose radiation effectiveness will be, as defined by the linear (α) coefficient. Thus, high energy gamma rays will have a low effectiveness and low-dose risk compared to X-rays and, within the different X-ray energies, the lower energy X-rays, for example mammography X-rays, will have a larger low-dose effectiveness and risk. It needs to be borne in mind that this is counter-intuitive. Softer X-rays might be expected to have lower risk than harder X-rays, which is not the case.

There is clearly no good reason to anticipate that the low-dose risk and thus radiation weighting factor of all sparsely ionising radiations will be the same and variations of up to a factor 5 seem quite credible. The use of different radiation weighting factors for different sparsely ionising radiations will adversely affect the use of dose equivalent and seriously complicate radiological protection practice. Therefore, the value of the radiation weighting factor for sparsely ionising radiations needs to be seriously reviewed.

In conclusion, the three important concepts which form the basis of internationally recommended radiological protection regulations have been shown to be without any valid scientific basis and cast serious doubt on the current (2018) defined radiological risk level.

The ICRP and other regulatory bodies need to review the whole basis for the determination of low-dose risk. The model presented in this book can provide a way forward with a similar philosophy of low dose and low-dose rate radiation action which is linear from the origin up but which is not associated with the 'linear no-threshold concept' that has been used until now to derive low-dose radiation risk. In addition, there are two epidemiological studies of radiological workers (Leuraud et al. 2015; Richardson et al. 2015) which provide low-dose risk data as a base for further deliberations.

These last three sections should not be taken in any way as criticism of the ICRP and other regulatory bodies. It is merely a call for a serious revision of radiological protection risk values.

9.5 A POTENTIAL WAY FORWARD

An important virtue of the model developed in this book is that it is based on a molecular mechanism, the induction of DNA double strand breaks by ionising radiation. The mechanism proffers equations for the induction of somatic mutations, chromosomal aberrations and cell inactivation and, by extension, the equation for cancer induction. The full equations are linear–quadratic with radiation dose and it has been shown that at low-dose rate these equations reduce, as a consequence of the perfect repair of DNA single strand breaks during chronic exposure, to a linear form with dose. This means that the risk from low dose and low-dose rate radiation exposure will increase in proportion with the exposure dose from zero dose up. This linear increase is compatible with the current postulate of the regulatory bodies but is quite different from the 'linear no-threshold' concept used by them. While the 'linear no-threshold' concept will provide virtually no information about the low dose, low-dose rate radiation risk, the linear (α) coefficient derived through the model is of direct relevance to that risk.

The model permits a direct approach to the problems highlighted in Sections 9.1 and 9.3. Although Equation 9.1 can be used to fit the full dose range of cancer induction following an acute exposure, the fitting will only provide inaccurate values of the linear (α) coefficient which is the coefficient of most relevance to the low dose, low-dose rate effect. Only data on the induction of cancer in man following chronic radiation exposures will provide good information on the low-dose, low-dose rate radiation risk. It is fortunate that there are two recent publications which deal precisely with the induction of cancer after chronic lifetime exposures: that by Leuraud et al. (2015) dealing with leukaemia and lymphoma in radiation workers and that by Richardson et al. (2015) dealing with cancer other than leukaemia. Both of these studies show a linear response for the excess relative risk (ERR) for cancer induction with accumulated exposure in line with the expectations of the model. Unfortunately, the UNSCEAR epidemiological study (UNSCEAR 2018) of cancer risk in populations exposed to environmental sources of radiation was inconclusive.

The two studies by Leuraud et al. (2015) and Richardson et al. (2015) should provide a strong basis for the future development of chronic radiation risk. However, the excess relative risk values presented by both Leuraud et al. and Richardson et al. quote the values as a function of dose in Gray and not dose equivalent in Sievert, raising the question of which weighting factor to use to convert these risks into Sievert. This is especially difficult because, if a weighting factor of 1 is used, the risks implied by the studies are considerably greater than that recommended by the regulatory bodies and if a factor greater than 1 is applied, other problems arise. The studies, while offering the best approach to deriving chronic radiation risk raise two additional problems.

The first concerns the appropriate weighting factor for the radiations experienced by radiation workers which can be addressed by unravelling the variety of these radiations in the INWORKS studies (Leuraud et al. 2015; Richardson et al. 2015). In addition, careful cellular radiobiological studies, using acute exposures of a number of different sparsely ionising radiations to determine accurate values of the linear (α) coefficients for cell inactivation, chromosomal aberrations and somatic mutations, could provide supporting information. Radiobiological studies will give a good idea of the range of the linear (α) coefficients which may be encountered in sparsely ionising radiation exposures and, consequently, what range of weighting factors may need to be implemented.

The second problem concerns the mechanism by which radiation induces cancer via a somatic mutation and, although the range of linear (α) coefficients determined in the radiobiological studies will provide a good starting point for the derivation of chronic radiation risk using Equation 9.1, it has to be realised that this equation is only approximate and valid for short-term acute and chronic exposures. It will be necessary to return to the full Equation 9.2 for cancer induction (see Chapter 7) which carries a value of the number of cells at risk, probably stem cells or intermediate cells (see Section 9.5), and to consider how this parameter might change with age and affect the cancer induction risk over the lifetime:

$$CI = 1 - \exp\left(-m\left(\vartheta\left(1 - \exp\left(-q\left(\alpha D + \beta D^2\right)\right)\right) \cdot \exp\left(-p\left(\alpha D + \beta D^2\right)\right)\right)\right) \quad (9.2)$$

where (m) is the probability that a malignantly mutated cell grows into a tumour and (ϑ) is the number of cells at risk for the malignant mutation per organism.

9.6 INSIGHTS INTO RADIOLOGICAL PROTECTION RISK FROM CANCER MODELLING

Cancer is a disease of old age and, almost certainly, arises from the accumulation of critical mutations in stem cells and their progeny over the lifetime. It is clear from the many molecular biological studies of various cancers (Stratton et al. 2009; Cancer Genome Atlas Research Network 2017a,b; Scarpa et al. 2017) that somatic mutations are an important event in the induction of cancer but these studies invariably find many different mutations in a single tumour. The question that arises is, 'How many critical mutations are necessary in a single stem cell to convert that cell to a malignant phenotype?'

Although the multi-stage nature of cancer had been discussed in the 1950s (Armitage and Doll 1954; Burch 1960), the seminal work of Knudsen (1971, 1985, 1991) on the occurrence of retinoblastoma in children led him to propose a two-mutation pathway to cancer development. Knudsen concluded from his studies that some retinoblastomas occurred in children who had inherited a mutation which was carried in all their cells and predisposed these children to the cancer induced by a single mutagenic event. Other sufferers were not genetically predisposed but developed the cancer as a result of two mutational events (Knudsen 1971). Together with Moolgavkar, Knudsen developed a two-stage model of carcinogenesis (Moolgavkar and Knudsen 1981), that identifies two mutational events which convert a normal stem cell to a malignant cell that ultimately develops into a tumour (see Chapter 7).

Figure 9.5 presents a modified version of the Moolgavkar and Knudsen model (1981) which is combined with the cellular model of radiation action developed in this book using the radiation-induced DNA double strand break as both a potential mutational event and as a cell inactivation event.

The radiation model for cellular effects has been combined with this two-mutation cancer model because it presents interesting and important perspectives for radiological protection regulation (Leenhouts and Chadwick 1994a).

The two-mutation model of cancer implies that 'spontaneous' cancers arise from 'spontaneous' mutations from the normal stage to the intermediate stage, $(\mu_1(0))$, and from the intermediate stage to the malignant stage, $(\mu_2(0))$. The development

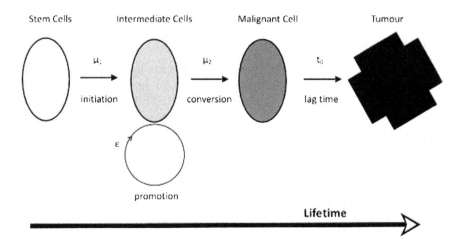

FIGURE 9.5 A schematic representation of a two-mutation (μ_1 and μ_2) model of cancer development containing the familiar stages of initiation (μ_1), promotion (ε), the ongoing division of initiated cells, conversion (μ_2) and lag time (t_0), which is the time for a single malignant cell to grow into a detectable tumour (note that the lag time is not the same as latency).

of cancer with lifetime using the two-mutation model is a continuous process with time which is difficult to define in an explicit mathematical equation. However, iterative and cumulative computations of the numbers of cells in the different stages can be used if approximate starting estimates are made of, for example, the number of normal stem cells at risk and the growth of that number in youth (Leenhouts and Chadwick 1994a). Moolgavkar (1983) has shown that the two-mutation model can describe the age-dependent increase in spontaneous cancer incidence in different cancers. Radiation exposure is assumed to be able to affect both mutational steps (μ_1 and μ_2) and it is also assumed that radiation can influence the ability of the intermediate cells to survive and divide. In addition, the division rate of the normal and intermediate cells is assumed to remain the same and be unaffected by either the radiation or the first mutation. It is important to realise that, although the original population of normal stem cells may be quite large, after the first mutation (μ_1), be it by a 'spontaneous' mutation or from a radiation-induced mutation, the number of intermediate cells is 1. The intermediate population of these mutated cells will grow exponentially by cell division. This feature will hold at each mutational step on the pathway to cancer even if there are more than two mutations in that path. It also holds for the mutation which converts an intermediate cell to a malignant phenotype. There will only be one malignant cell to start the tumour (Greaves and Maley 2012).

The combined model can be used to calculate simultaneously the age and radiation dose dependence of tumour development. Two different exposure scenarios must be considered: an acute or chronic short-term exposure within the period of a cell cycle and a long-term exposure to very low-dose rate radiation over a large part of lifetime. In the first short-term exposure situation, the defining radiation quantity for use in the cellular model is the total dose of radiation but in a long-term exposure to very low-dose rate over a large part of lifetime, the radiation quantity for use in the cellular model is the dose per cell cycle.

9.6.1   SHORT-TERM EXPOSURE

A short-term exposure is clearly at a particular moment in the lifetime and, according to Figure 9.5, will affect either the initiation stage, ($\mu_1(0)$), or the conversion stage, ($\mu_2(0)$), of the two-mutation cancer model. It is unlikely to affect both stages. If the short-term exposure occurs at a young age there will be few spontaneously induced intermediate cells and the radiation will mainly affect the initiation stage. The probability for the mutation of a stem cell to an intermediate stage will, momentarily, be defined by

$$\mu_1(D) = \mu_1(0) + \left(1 - \exp\left(-q\left(\alpha D + \beta D^2\right)\right)\right) \cdot \exp\left(-p\left(\alpha D + \beta D^2\right)\right) \qquad (9.3)$$

for an acute short-term exposure and

$$\mu_1(D) = \mu_1(0) + \left(1 - \exp\left(-q\left(\alpha D\right)\right)\right) \cdot \exp\left(-p\left(\alpha D\right)\right) \qquad (9.4)$$

for a chronic short-term exposure.

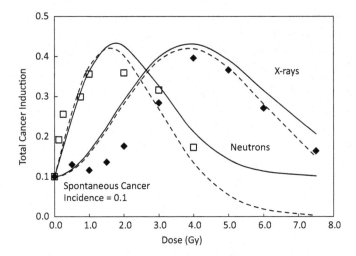

FIGURE 9.6 The induction of lung cancer in female mice after X-ray (closed diamonds) and neutron (open squares) exposures superimposed on the spontaneous cancer induction of 0.1 (data from Coggle 1988). The dotted lines show the potential effect of cell killing on the cells involved in the spontaneous induction of cancer.

In other words, the dose–effect relationship for a short-term exposure will be superimposed on the spontaneous mutation rate and thus on the spontaneous cancer induction (see Figure 9.6).

A similar situation will arise if the short-term exposure occurs later in life, when the population of spontaneously induced intermediate cells will have expanded exponentially with time, and the radiation affects the conversion stage. The probability for the mutation of an intermediate cell to a malignant phenotype will, momentarily, be defined by

$$\mu_2(D) = \mu_2(0) + \left(1 - \exp\left(-q\left(\alpha D + \beta D^2\right)\right)\right) \cdot \exp\left(-p\left(\alpha D + \beta D^2\right)\right) \qquad (9.5)$$

for an acute short-term exposure and

$$\mu_2(D) = \mu_2(0) + \left(1 - \exp\left(-q(\alpha D)\right)\right) \cdot \exp\left(-p(\alpha D)\right) \qquad (9.6)$$

for a chronic short-term exposure.

It has to be borne in mind that, at the higher doses when cell killing starts to dominate the cancer induction relationships outlined in Equations 9.3 to 9.6, cells involved in the spontaneous induction of cancer will also be killed and the tail-off of the cancer induction curve at higher doses can be expected to fall somewhere between the solid curves and the dotted curves shown in Figure 9.6. Fortunately, the rising portions of both curves shown in Figure 9.6, which indicate the induction of cancer by the radiation, hardly differ and are not really influenced by cell killing.

This explains why the dose–effect relationships found after acute exposures of, for example, animal models, can be analysed using a mutation type of equation and why the known dependence on dose rate and radiation quality are found (see Figures 7.3 to 7.6 in Chapter 7).

Two important conclusions can be drawn. The first is that at low doses and low-dose rates, the radiation induction of cancer is linear with dose from zero dose up. It should be noted that, although this is not the same concept as the 'linear no-threshold' concept, it still means that there is absolutely no threshold dose for the induction of cancer.

Second, the radiation-induced mutation of either the stem cells or the intermediate cells implicitly relies on 'spontaneous' mutations to complete the process to malignancy. A stem cell induced by radiation to an intermediate stage relies on a 'spontaneous' mutation to convert it to a malignant phenotype. A stem cell 'spontaneously' mutated to an intermediate stage can be converted to a malignant phenotype by radiation. In other words, the radiation effect cannot be separated from the spontaneous tumour incidence which influences the level of the radiation effect. This interplay, between the spontaneous mutations and the radiation-induced mutations, results in the prediction that a higher level of spontaneous tumour incidence, presumably arising from a higher spontaneous mutation frequency, will lead to a steeper initial slope of the radiation effect (see Figure 9.7).

However, differences in the number of stem cells in different organs could also influence the level of spontaneous tumour incidence and confuse this prediction. It is, nevertheless, a factor which needs to be taken into account in radiological risk definition.

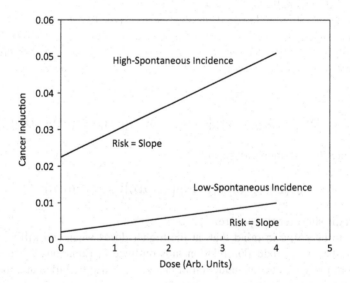

FIGURE 9.7 The result of a calculation using the cancer model of low-dose cancer induction risk at the end of life after an acute exposure at 20% of life. The two incidence curves show a low risk for a low-spontaneous cancer induction and a higher risk for a high-spontaneous cancer induction. The number of stem cells, the division rate of the intermediate cells and the radiation sensitivity of the cells were held constant in the calculations. Only the spontaneous mutation rates were altered.

This interplay between a 'spontaneous' mutation and a radiation-induced mutation in the development of cancer following a short-term exposure suggests that radiation is a co-factor in cancer induction. A similar argument can be made for the short-term exposure to any other mutagenic agent.

Interestingly, a short-term exposure at a young age when radiation is more likely to mutate a stem cell to an intermediate stage leads to an age-dependent tumour incidence which is comparable with a relative risk model. A short-term exposure at later ages, when radiation is more likely to convert an intermediate cell to malignancy, leads to an age-dependent tumour incidence which is comparable with an absolute risk model (Leenhouts and Chadwick 1994a; Chadwick and Leenhouts 1995). This is illustrated in Figure 9.8.

This short-term exposure situation is relevant to the atomic bomb survivors and to most animal experiments.

9.6.2   Long-Term Exposure

It should be clear, from the life long timeline in the two-mutation cancer model, that this long-term exposure situation, which is relevant for radiation workers, uranium miners and possibly the general public, is different from the short-term exposure. The short-term exposure will cause a momentary augmentation, related to the total exposure dose (see Equations 9.3 to 9.6), of the spontaneous mutation rate of one or other of the mutational steps and will affect the survival of the intermediate cells

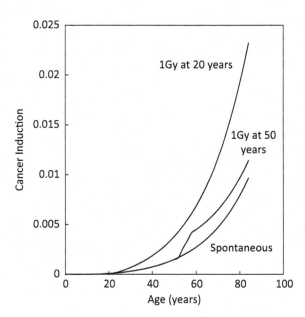

FIGURE 9.8   The simulated age-dependent increase in cancer induction using the combined two-mutation model following an acute exposure at 20 years of life which gives a relative type of risk compared with an acute exposure at 50 years of life which gives an absolute type of risk.

already present from the 'spontaneous' process. The long-term exposure will be an exposure extended over a considerable period of life and the radiation will add a more or less continuous contribution to the 'spontaneous' mutation rates in both the initiation stage and in the conversion stage of the cancer process. The important question which needs to be investigated is whether long-term exposure leads to any changes, such as augmentation of cell division, possibly resulting from ongoing promotion during lifetime, which would alter the radiological risk derived for a short-term chronic exposure.

The model calculations to examine low-dose radiation risk for long-term exposure can be made using the dose per cell cycle (D_c).

If $(\mu_1(0))$ and $(\mu_2(0))$ are the cancer model coefficients describing the probabilities for the two 'spontaneous' mutations without radiation, then at a long-term exposure dose of (D_c) per cell cycle these coefficients become

$$\mu_1(D_c) = (\mu_1(0) + q_1 D_c) \cdot \exp(-(pD_c)) \tag{9.7}$$

and

$$\mu_2(D_c) = (\mu_2(0) + q_2 D_c) \cdot \exp(-(pD_c)). \tag{9.8}$$

It is quite rational to assume that the repair of single strand breaks will be complete during the very long-term exposures so that $\beta = 0$. It is also rational to assume that the cell division rate of the stem cells and the intermediate cells, $(\varepsilon(0))$, is not altered by the radiation or by the mutation to the intermediate cell stage. Cell killing by radiation during a long-term exposure, which is included in Equations 9.7 and 9.8, is unlikely to have any significant effect on the cancer induction process because of the low doses per cell cycle normally involved.

When the spontaneous cancer prevalence and, consequently, the 'spontaneous' mutation rates, $(\mu_1(0))$ and $(\mu_2(0))$, are non-negligible and the 'spontaneous' mutation rate is larger than the radiation-induced mutation rate per cell cycle, it is probable that a radiation-induced mutation in one step will interact with a 'spontaneous' mutation in the other step of the cancer process. In this case, the dose–effect relationship for radiation risk at low doses will be proportional with total accumulated dose. There will not be a threshold dose. This is illustrated in Figure 9.9 and radiation can again be considered as a co-factor in the induction of cancer. When the spontaneous cancer prevalence is very small, and the 'spontaneous' mutation rates are almost negligible, the radiation-induced mutation rate per cell cycle could dominate. In this case, a radiation-induced mutation of a stem cell could, at a later time, be converted to a malignant phenotype by a second radiation-induced mutation, and the low-dose radiation risk could show signs of a quadratic component. This is also illustrated in Figure 9.9 and, in this case, radiation can be considered as a complete carcinogen. It is important to note that, even in this case, the very low-dose risk will still be linear with the dose although the slope will be very small. There will not be a threshold dose below which cancer is not induced.

A specific example of the quadratic nature of radiation-induced cancer risk is found in bone cancers induced in the radium dial painters. The spontaneous incidence

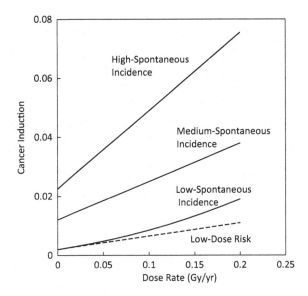

FIGURE 9.9 The result of a calculation using the cancer model of low-dose cancer induction risk at the end of life after a long-term exposure over a large period of life for a low-spontaneous cancer induction, a medium-spontaneous cancer induction and a higher-spontaneous cancer induction. The number of stem cells, the division rate of the intermediate cells and the radiation sensitivity of the cells were held constant in the calculations. Only the spontaneous mutation rates were altered. The higher-spontaneous incidence cancer shows a higher risk. At the low-spontaneous cancer induction there is a hint of a quadratic component.

of primary bone cancer is very low and the radiation dose to the bones of the dial painters was from alpha particle emitters so that the induction of mutations can be assumed to be proportional with alpha particle dose.

The induction of bone cancers in the dial painters can be described using the combined two-mutation model as illustrated in Figure 9.10. These bone cancer data in female dial painters are often used to claim that there is a threshold dose for cancer induction but this is not the case, even though the risk at very low doses is very small (Leenhouts and Brugmans 2000). At very low doses, there is a linear component of risk which arises from the interaction of a very low 'spontaneous' mutation with an alpha particle-induced mutation in the pathway to bone malignancy. At higher doses, the malignancy is induced by two alpha particle-induced mutations, thus giving a quadratic component to the risk.

The analysis of the radium dial painters by Leenhouts and Brugmans (2000) demonstrates that, even though radiation can often be interpreted as a co-carcinogen, it can also be a complete carcinogen. The complete carcinogenic action of radiation will only be seen in long-term exposures and in cancers which show a very low natural incidence.

The data shown in Figure 9.10 at the highest radium intakes reveal a cumulative cancer incidence of 100% even though the number of cases is small and the errors are substantial. It has been shown that, in short-term exposures, the incidence of

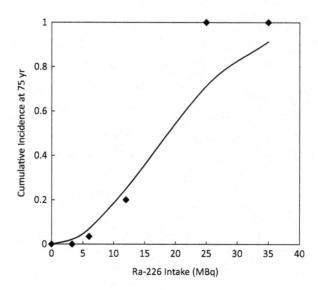

FIGURE 9.10 The incidence of bone cancer in female dial painters as a function of radium-226 intake described by a quadratic function of intake (data from Rowland 1994; redrawn from Chadwick et al. 2002).

cancer peaks at around 30% in animal studies and decreases at higher doses, as a result of cell killing. In a long-term exposure to very high doses, it appears that the combination of two radiation-induced mutations, coupled with the expansion of intermediate cells, can overcome the cell-killing effects. This potential effect of long-term exposure needs to be taken into consideration in the future review of radiological risk.

The interplay between the radiation-induced mutations and the 'spontaneous' mutations seen after both a short-term exposure and, although somewhat differently, after a long-term exposure, means that the spontaneous cancer incidence in a population plays a role in defining the radiation-induced risk. Spontaneous cancer incidence varies in different population groups and the model indicates that risks might also differ in different population groups but it offers, nevertheless, a potential pathway to transfer risk from one population group to another.

One other feature arising from the model is that, as cancer develops from the accumulation of 'spontaneous' mutations, people who carry a gene in their cells which affects DNA repair will accumulate the mutations more rapidly and will, consequently, develop cancer at a much earlier age, for example, in young adulthood (25–55 years). The radiation aspect of the model implies that people carrying a gene giving defective repair of DNA double strand breaks, for example a BRCA gene, will not only develop cancer in young adulthood but will also be extra sensitive to any radiation therapy treatments. A similar, more general, sensitivity could also exist for chemotherapy treatments of young adults with cancer.

Clearly, both short-term and long-term exposure situations are relevant for radiological protection and the determination of radiological risk but the use of the combined radiation-cancer model defines two different approaches with potentially different results for radiological risk assessment. Although the acute data on cancer induction in man is unlikely to provide much information on the chronic radiation risk, the acute data may provide other information which could be of relevance to low-dose radiation risk, such as the variation of sensitivity to cancer induction with age. The studies of the radiation workers published by Leuraud et al. (2015) and Richardson et al. (2015) would be the right place to gain information for the risk from a long-term exposure.

It is important to note that, in general, even if it is eventually shown that three or four mutations are required in the chain taking a normal stem cell through the multi-step process to a malignant cell, the perspectives offered for radiological risk assessment by the combined two-mutation model do not change much.

Only the most interesting perspectives arising from the use of the combined two-mutation model for radiological protection have been briefly discussed here and the interested reader is referred to the following series of publications which delve more deeply into the use of the combined radiation-cancer model (Leenhouts 1999; Leenhouts and Chadwick 1994a,b, 1997; Leenhouts and Brugmans 2000; Leenhouts et al. 1996, 2000; Venema et al. 1995; Chadwick et al. 1995, 2002).

In conclusion, the perspectives offered by the combined two-mutation model of cancer induction for radiological risk assessment suggest that the regulatory bodies have some work to do. The articles by Leuraud et al. (2015) and Richardson et al. (2015) on the risk of cancer determined in radiation worker populations exposed over a large part of the lifetime would appear to be a good place to start.

9.7   HEREDITARY EFFECTS

The 'linear no-threshold' concept was, in fact, originally derived from the work of Muller (1928) on hereditary effects and taken up later by the regulatory bodies for the analysis of cancer induction in the atomic bomb survivors. Calabrese (2017) has recently done a clinical job of raising doubts about the validity of the original work on hereditary effects, the specificity of the mutants scored and the value of the 'linear no-threshold' concept derived from the work of Muller.

Using the DNA double strand break model, additional comments can be made. In the first place, the mutant data measured by Muller can be expected to follow a dose–effect relationship defined by Equation 9.1 and it is, therefore, not surprising that Muller interpreted it as a linear dose–effect curve through the origin, thus initiating the 'linear no-threshold' concept (see Figure 9.3).

Second, in view of the work of Thacker et al. (1977), Rao and Hopwood (1982) and Iliakis (1984a,b) (see Chapter 3, Figures 3.12 to 3.14), the mutation rate for specific mutations (HGPRT) is of the order of 1 in 10^5 surviving cells, so the mutants scored by Muller are hardly likely to have been very specific.

Third, in view of this very low induction of specific mutations at mitosis it seems improbable that, in spite of the potential number of target cells, a noticeable increase in specific hereditary effects would be found in the offspring of atomic bomb survivors.

Finally, it can be predicted that. in accordance with the model based on DNA double strand breaks, the dose–effect for hereditary mutations will be linear–quadratic after an acute exposure reducing to a linear effect at low-dose rate chronic exposure and the different effects of different radiation quality generally found in somatic cellular studies will be reproduced in hereditary mutations.

It is thus concluded that the risk for hereditary effects will be quite small.

9.8 CONCLUSIONS

The model equation derived to fit cancer induction reveals that the fitting is dominated by the quadratic (β) coefficient and gives no information about low-dose risk. This means that the straight line fitting to the rising cancer data, derived from the 'linear no-threshold' concept used by regulatory authorities, also gives no information about low-dose risk.

It is concluded that the 'linear no-threshold' (LNT) concept and accompanying 'dose and dose rate reduction factor' are not useful for radiological protection and should be discarded. The weighting factor of 1 for all sparsely ionising radiation, currently used by regulatory bodies, is also cast into doubt.

The regulatory bodies need to revise their approach to the derivation of low-dose risk.

Part II

Ultraviolet Light Effects

10 The Molecular Model and the Cytotoxic Action of UV Light

The important role for oxidative damage in the UVA effect implies that pairs of oxidative DNA single stranded damage form the critical lesion and not pyrimidine dimers. Cell killing by UVA light can be analysed using a purely quadratic exposure equation. A comparison is made between the formation of the paired lesion induced by UVA and the ionising radiation-induced double strand break formed in the beta mode of radiation action. The purely quadratic nature of the induced cell killing by UVA suggests that, at very low UVA exposure levels, there will be little health risk. The model developed for ionising radiation is extended to the effect of UVB and UVC light, taking advantage of the suggestion by Park and Cleaver that, two pyrimidine dimers on opposite DNA strands and on either side of a replication terminus could block replication and be a potentially lethal lesion. This suggestion leads to a purely quadratic exposure response equation for cell killing and also mutation induction which is borne out by the analysis of experimental data. The two-dimer lesion is only recognised in DNA synthesis and the cells cease to divide in the replication phase, in contrast to the situation after ionising radiation. The recognition of the two-dimer lesion at replication means that each individual dimer can be repaired at any time during and after exposure until DNA replication starts. Thus, in stationary cells with delayed plating, a complete and correct repair of the UV-induced damage can be achieved. This is also borne out by experimental data. It is suggested that the cells blocked in replication by the two-dimer lesion might be in a state of senescence.

10.1 THE SPECTRUM OF ULTRAVIOLET LIGHT

Ultraviolet light falls between visible light and X-rays and gamma rays on the spectrum of electromagnetic radiation and is strongly associated with skin cancer (Kraemer 1997). The UV range is usually divided up into UVA (320–400 nm), UVB (280–320 nm) and UVC (200–280 nm). UVA, which borders violet in the visible spectrum and is some 30 times more abundant than UVB in sunlight, was generally thought to be relatively harmless at low exposures although there are reports of single strand oxidative damage to DNA (Cadet et al. 2015) and an association with skin cancer induction (Buckel et al. 2006; Ting et al. 2007; Moan et al. 2008). UVB and UVC photons can cause single strand DNA damage in the form of pyrimidine dimers. However, UVC photons (200–280 nm), which are absorbed by the atmosphere, have enough photon energy (4.5–12 eV) to be slightly ionising and capable of inducing single strand breaks in DNA although they cannot induce DNA double strand breaks in the alpha mode of radiation action. The biological action of both the DNA single strand breaks induced by UVA and UVB and the pyrimidine dimers induced by UVB and UVC, remains to be elucidated and similarities and differences with the action of ionising radiation need to be identified.

10.2   THE CELL-KILLING EFFECT OF UVA LIGHT

The review by Cadet et al. (2015) reveals a wide range of photoproducts induced in the cell by UVA exposure and also reveals the importance of the oxidative damage, such as DNA single strand breaks and 8-oxo-7,8-dihydroguanine DNA base damage, for the biological effect. However, there are also contradictory reports for pyrimidine dimers as well as DNA double strand breaks, chromosomal aberrations, and mutations induced by UVA (Peak and Peak 1990a; Fell et al. 2002; Rapp et al. 2004; Wischermann et al. 2008; Greinert et al. 2012; Fang et al. 2014; Rizzo et al. 2011; Douki et al. 2003; Mouret et al. 2006; 2010). The carcinogenic potential of UVA has also been reported with some emphasis on the induction of melanoma (Buckel et al. 2006; Ting et al. 2007; Moan et al. 2008).

In relation to cell killing, Peak et al. (1991a) have demonstrated that the induction of DNA single strand breaks in Chinese hamster ovary cells is proportional with 365 nm UVA exposure and that the cell survival relationship after 365 nm UVA exposure is curved. Figure 10.1a presents the survival of a H_2O_2 resistant Chinese hamster cell line as a function of 365 nm UVA exposure and Figure 10.1b shows the same data but cell survival is plotted as a function of the square of 365 nm UVA exposure. The straight line fitting in Figure 10.1b reveals that cell survival is directly related to the square of the UVA exposure. This suggests that pairs of UVA-induced photoproducts

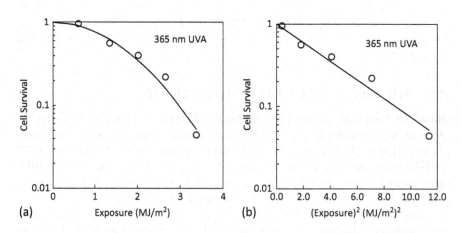

FIGURE 10.1   (a) The effect of 365 nm UVA exposure on cell survival in OC5 Chinese hamster ovary (CHO) cells as a function of exposure. (b) The same data as in (a) but with cell survival plotted as a function of the square of the UVA exposure (data from Peak et al. 1991a).

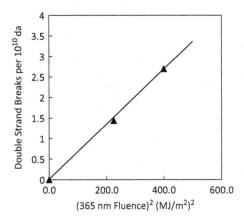

FIGURE 10.2 The number of DNA double strand breaks drawn as a function of the square of the 365 nm fluence. The straight line through the origin implies that each double strand break arises from two single independently induced photoproducts, probably single strand breaks (data from Peak and Peak 1990a).

form the critical lethal lesion which is strengthened by the cell survival studies after UVA exposure shown in Figures 10.3 and 10.4.

The analysis says very little about the nature of the UVA-induced 'paired lesion' but the suggestion by Cadet et al (2015) that the presence of oxygen during exposure plays an important role in the UVA effect might be taken to indicate that the 'paired lesion' arises from DNA single strand breaks or 8-oxo-7,8-dihydroguanine DNA damage or combinations of them and not from paired dimer lesions. The yield of DNA pyrimidine dimer damage is not influenced by the presence of oxygen during exposure (Cadet et al. 2015). If the UVA-induced 'paired lesion' does indeed arise from combinations of single strand breaks and 8-oxo-7,8-dihydroguanine base damage, then the process by which these combinations form needs to be elucidated.

Interestingly, Peak and Peak (1990a) have measured the induction of DNA double strand breaks by different wavelengths of UV light, including 365 nm UVA light, in cultured human P3 epithelioid cells and shown that the number of DNA double strand breaks measured using neutral filter elution was directly proportional to the square of the UVA exposure (see Figure 10.2). This suggests that a pair of single strand breaks forms one double strand break.

Drawing an analogy with the model for ionising radiation, a double strand break would be classed as a potentially lethal lesion which would explain the cell survival data shown in Figures 10.1b, 10.3b and 10.4b.

The formalism for the cellular action of UVA light exposure can be developed as follows: if (ε_A) is the probability per cell per unit exposure squared (X_A^2) that two independently induced photoproducts (single strand breaks or 8-oxo-7,8-dihydroguanine DNA base damage) form a DNA double strand break, then the number (N_A) of DNA double strand breaks is given by

$$N_A = \varepsilon_A X_A^2 \tag{10.1}$$

and cell survival is given by

$$S_A = \exp\left(-p\left(\varepsilon_A X_A^2\right)\right). \tag{10.2}$$

Although the survival curves shown in Figures 10.1a, 10.3a and 10.4a resemble the survival curves shown earlier following ionising radiation exposure, it is important

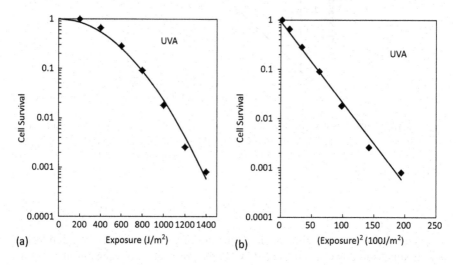

FIGURE 10.3   (a) The effect of UVA on cell survival in V22 Chinese hamster cells as a function of exposure. (b) The same data as in (a) but with cell survival plotted as a function of the square of the UVA exposure (data from Ikebuchi et al. 1988).

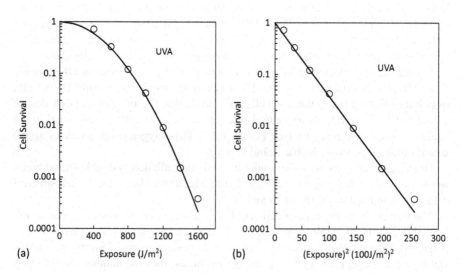

FIGURE 10.4   (a) The effect of UVA on cell survival in M29 cells as a function of exposure. (b) The same data as in (a) but with cell survival plotted as a function of the square of the UVA exposure (data from Ikebuchi et al. 1988).

to realise that there is absolutely no linear ($p\alpha$) coefficient in the case of the UVA-induced cell killing. The straight lines in Figures 10.1b, 10.3b and 10.4, passing directly through the origin, emphasise that point.

In their article studying UV light-induced DNA double strand breaks using neutral filter elution, Peak and Peak (1990a) showed that all wavelengths of UV light studied, from UVC to UVA, induced DNA double strand breaks. However, when the number of double strand breaks per lethal event was compared with that of ionising radiation, they found that only UVA wavelengths were closely comparable with the ionising radiation (eg, 365 nm UVA-induced 4.3 dsb/genome/D_0 and gamma radiation-induced 1.2 dsb/genome/D_0). The comparable cell-killing efficiency of DNA double strand breaks, induced by both ionising radiation and UVA light, extends the analogy between the two types of radiation. This analogy is further emphasised by the report that scavenging of hydroxyl radicals protects cellular DNA against breakage after ionising radiation and 365 nm UVA exposure (Peak and Peak 1990b). However, Peak and Peak (1990a) concluded that there were too few single strand breaks for the double strand breaks to arise from two closely opposed single strand breaks.

This is the same sort of problem encountered with the formation of DNA double strand breaks from two independently induced single strand breaks in the beta mode of ionising radiation action. In Chapter 2, Section 2.5, it was concluded that two different processes were involved in the formation of ionising radiation induced double strand breaks in the beta mode of radiation action. This conclusion was based on results showing different scavenging effects on the alpha mode and beta mode of radiation action as well as information from radiation quality analyses. In the beta mode of radiation action, the first process involved a radical, probably a hydroxyl radical, formed very close to the DNA helix (0.7 nm) which caused a single strand break. This break was proposed to make the remaining intact single strand vulnerable to a second event occurring some distance from the DNA (8 nm) which converted the first single strand break into a double strand break. It would be interesting to know if there are similarities between the formation of the UVA-induced 'paired lesion' and the ionising radiation-induced DNA double strand breaks formed via the beta mode of radiation action. A question that remains is whether the 'second' photoproduct converts the 'first' single-strand break directly into a double strand break or not. If it does and the UVA-induced 'paired lesion' is formed by a similar process to the ionising radiation beta mode lesion giving a frank DNA double strand break, then the reported induction of DNA double strand breaks would be explained and that would account for the findings of chromosome aberrations and mutations and also the indications of cancer (Fell et al. 2002; Raap et al. 2004; Wischermann et al. 2008; Greinert et al. 2012; Fang et al. 2014; Buckel et al. 2006; Ting et al. 2007; Moan et al. 2008).

Bearing in mind the association made in Chapter 4 between a double strand break and a chromosome aberration, it is regrettable that no publications have been found which document a relationship between chromosome aberration yield and UVA exposure. The anticipated quadratic exposure relationship, if verified, would provide convincing evidence in favour of the proposal that a chromosome aberration arises from one double strand break (Chadwick and Leenhouts 1978, 1981).

It is important to realise that there is no linear ($p\alpha$) coefficient in the case of the UVA-induced cell killing. The purely quadratic nature of the induced cell killing by UVA suggests that at very low UVA exposure levels there will be little health risk, although it must be remembered that the repair of DNA double strand breaks is never perfect. Clearly, attention should be given to the regulated use of sun beds for tanning purposes and to the need for advice about sunbathing practices. In view of the relatively rapid repair of both DNA single strand breaks (+/– 40 mins, Dugle and Gillespie, 1975) and 8-oxo-7,8-dihydroguanine (+/– 30 mins, Beseratinia et al. 2008) investigation into the potential reduction of the effect of UVA at lower exposure rates seems warranted. However, it should be kept in mind that these repair rates are measured in metabolically active cells and not the stationary or more slowly dividing skin cells where repair could be slower. In addition, the increased effect of UVA on cytotoxicity when the exposure rate is reduced (McMillan et al. 2008; Shorrocks et al. 2008), a somewhat counter-intuitive result, needs to be resolved.

10.3   THE MOLECULAR MODEL FOR THE ACTION OF UVB AND UVC LIGHT

In contrast to the UVA results reported above, the number of double strand breaks per lethal event induced by UVB and UVC exposure was 100 times smaller than that for UVA and gamma radiation (Peak and Peak 1990a), indicating that a different photoproduct was responsible for killing cells after UVB or UVC exposure. In addition, a comparison of cell survival curves after UVA and UVB exposure, reveals that the exposure levels required for a reduction of cell survival to 0.1% are much smaller for UVB than for UVA. UVB gives an exposure level of about 7 J m^{-2} while the exposure following UVA gives an exposure of about 1000 J m^{-2} (see Figure 10.3a). Even though the cells exposed to UVB are not Chinese hamster cells, this difference of a hundredfold in exposure levels is remarkably large.

UVB and UVC photons can induce deformational damage to a single strand of DNA, usually in the form of a pyrimidine dimer, where two thymidine nucleotide bases are adjacent to each other along one strand of the DNA. These pyrimidine dimers cause a deformation in the DNA strand and can be repaired using the undamaged DNA strand. This provides the complementary base sequence to guarantee the correct restoration of bases along the damaged strand at the deformation and gives perfect repair. This is similar to the perfect repair of DNA single strand breaks, induced by ionising radiation, but it takes longer.

In studying the repair of pyrimidine dimers at different stages of the cell cycle, Cleaver and his colleagues (Cleaver 1981; Cleaver et al. 1979; Park and Cleaver 1979a,b) came to the conclusion that UV-induced deformations (dimers) could block the progress of a DNA replication fork. They also suggested that this block of a replication fork could be relieved by a replication fork coming in the opposite direction if it continued past the normal replication terminus.

In Chapter 1, Section 1.2, the process of DNA replication in the DNA synthesis phase of the cell cycle was described. In DNA synthesis, the two 'old' strands of DNA loosen and replication starts at many replication origins in both directions along the DNA. Cleaver and his colleagues extended the concept of the dimer-blocked replication fork by proposing that, if two replication forks approaching each other were both blocked by two dimers on the opposite DNA strands and on either side of a

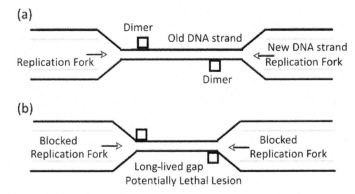

FIGURE 10.5 A schematic representation of the block to replication completion caused by the two pyrimidine dimers, one on each strand of the un-replicated DNA, giving a long-lived gap, a potentially lethal lesion.

replication terminus, a 'long-lived gap' would be created preventing the completion of DNA replication (Park and Cleaver 1979a). It was suggested that this 'long-lived gap' could be a potentially lethal lesion for the cell (see Figure 10.5).

This potentially lethal lesion consisting of two dimers, one on each strand of the un-replicated DNA, has basic similarities to the DNA double strand break discussed in Chapters 1 to 9 as the critical ionising radiation-induced lethal lesion. Consequently, the two-dimer lesion has been adopted as a molecular model for the critical UVB and UVC-induced cellular lesion and a quantitative approach to UVB and UVC-induced cellular effects has been developed.

10.4 A QUANTITATIVE APPROACH TO THE EFFECTS OF UVB AND UVC EXPOSURE

The formalism of the cellular action of UVB and UVC light, according to the proposal of Park and Cleaver (1979a), is developed from the following basic postulates:

- The number of single strand photoproducts (dimers) induced in the DNA molecule in the cell is strictly proportional to the exposure (X) of the UVB or UVC light.
- The single strand photoproducts (dimers) can be perfectly repaired, given time.
- The occurrence of a pair of photoproducts (dimers) on opposite strands of the DNA at two converging DNA replication forks in DNA synthesis can block the completion of replication and form a long-lived gap.
- The number of photoproduct pairs, or potentially critical lesions, is proportional to the square of the exposure (X^2) of the UVB or UVC light.
- The photoproduct pairs are only recognised as such at the moment of DNA replication (DNA synthesis).
- The individual photoproducts (dimers) in the pair can be repaired during and after exposure until replication occurs.

The number of pairs of photoproducts (dimers) (N_d) induced by an exposure (X) of UVB or UVC light is

$$N_d = \varepsilon_d X^2 \qquad (10.3)$$

where (ε_d) is the probability per cell per unit exposure squared that two independently induced photoproducts (dimers) form a pair (Chadwick and Leenhouts 1983).

Then, by analogy with the equations developed for the cellular effects of ionising radiation, cell survival (S_d) is given by

$$S_d = \exp\left(-p\left(\varepsilon_d X^2\right)\right). \qquad (10.4)$$

Two survival curves are presented in Figure 10.6. Figure 10.7 shows the same data plotted as a function of the square of the UVB exposure (X).

Although the survival curves shown in Figure 10.6 resemble the survival curves shown earlier following ionising radiation exposure, it is important to realise that there is absolutely no linear ($p\alpha$) coefficient in the case of the UVB-induced cell killing. The straight lines in Figure 10.7, passing directly through the origin, emphasise that point.

Some good experimental data supporting the 'paired dimer' lesion as critical for UV cellular effects are to be found in the work of Wade and Lohman (1980). They measured UV-induced cell survival in chicken embryo cells before and after photoreactivation and correlated survival with the number of endonuclease-sensitive sites,

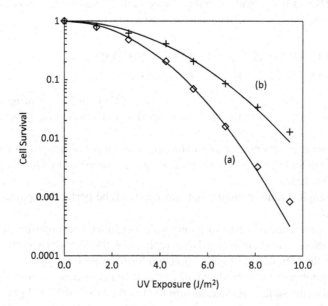

FIGURE 10.6 UV survival curves of (a) a homogeneous cell population of EAT cells and (b) a quasi-stationary population of EAT cells, drawn as a function of UV exposure (data from Iliakis and Nusse 1982). The lines are drawn in accordance with Equation 10.4 and are purely quadratic functions. The tail-out at the highest exposure is probably due to G2 phase cells.

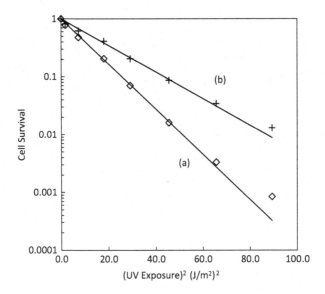

FIGURE 10.7   The same data as shown in Figure 10.6 but cell survival is drawn as a function of the square of UV exposure (data from Iliakis and Nusse 1982). The straight lines are drawn in accordance with Equation 10.4 and have the slope ($p\varepsilon_D$). The tail-out at the highest exposure is probably due to G2 phase cells which have a longer time, until the next DNA synthesis, to repair the pyrimidine dimers.

which are presumed to be dimers. Their correlation is shown in Figure 10.8 where cell survival is drawn as a function of the square of the number of endonuclease-sensitive sites (ESS). The single straight line correlation of the data down to 5% survival, with and without photo-reactivation, implies that 'paired dimers' are the critical lesion.

By analogy with the equations developed for the cellular effects of ionising radiation, mutation frequency is given, approximately, by

$$M_D \approx q\varepsilon_D X^2. \tag{10.5}$$

Figure 10.9 presents some data on 6-thioguanine resistance measured in human fibroblasts exposed one hour before the DNA synthesis phase by Konze-Thomas et al. (1982) analysed according to Equation 10.5.

It is important to differentiate between the effects of ionising radiation discussed in the first nine chapters and the UVB or UVC exposures discussed here.

- A single photon of UVB or UVC light cannot create a 'paired dimer' lesion in the DNA and the exposure relationships for UVB or UVC exposure will not exhibit a 'linear coefficient' as is the case with ionising radiation. This means that there will be very little risk of harm at very low-level UVB or UVC exposure.
- Although the UVB or UVC exposure relationship does have a purely quadratic ($p\varepsilon_D$) coefficient and appears similar to the quadratic ($p\beta$) coefficient for ionising radiation the 'paired dimer' lesion' and the 'DNA double strand break' are quite different lesions.

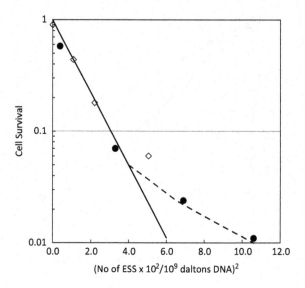

FIGURE 10.8 The correlation between cell survival, with photo-reactivation (open diamonds) and without photo-reactivation (closed circles) and the square of the number of endonuclease-sensitive sites (ESS), presumed dimers, measured in chicken embryo cells after UVB exposure (data from Wade and Lohman 1980). The correlation supports the 'paired dimer' lesion concept. The tail-out at higher exposures probably comes from resistant G_2 phase cells.

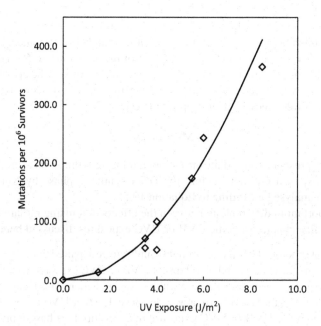

FIGURE 10.9 The mutation frequency (6 thioguanine resistance) induced in human fibroblasts one hour before the DNA synthesis phase by UV exposure analysed as a purely quadratic function of exposure in accordance with Equation 10.5 (data from Konze-Thomas et al. 1982).

- The ionising radiation-induced 'DNA double strand break' is created at the moment when the 'first' single strand break is converted to a double strand break by the 'second' single strand break, whereas the two independently induced dimers are only recognised as a 'paired dimer' lesion when the cell enters DNA synthesis after exposure and the replication fork is blocked.
- This means that the 'paired dimer' lesion is a virtual lesion until it is recognised at DNA synthesis and therefore each individual dimer has time to be repaired perfectly both during and after UVB or UVC exposure until the cell enters the DNA synthesis or S phase. Consequently, there will not be a dose rate effect for UVB or UVC exposure.
- It also means that cells exposed in the G_2 cell phase will have the time from exposure until the cell enters the next S phase to repair the dimers and will therefore be much more resistant than cells exposed in the G_1 cell phase immediately prior to the S phase.
- And this means that, in stationary cells, all the UVB or UVC-induced dimers can be repaired, given sufficient time, before the cells are released from the stationary phase and go into the cell cycle. The exposed cells will have no record of the exposure.

The consequence of this for sunbathing is that, as skin cells only divide slowly, after mild exposures of UVB the cells will be able to repair the dimers and there will be very little, or no cell killing and little or no indication of sun-burning.

In this respect, Figure 10.10 presents some data from Maher et al. (1979) showing the effect of UVB exposure on survival (Figure 10.10a) and mutation induction

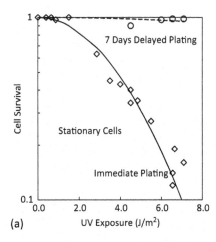
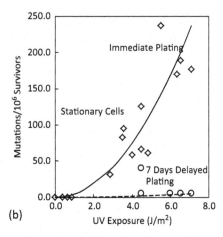

FIGURE 10.10   (a) The influence of a long repair time on the survival of stationary human fibroblast cells following UVB exposure. The immediate plating survival curve (open diamonds) is fitted with a purely quadratic equation. The long delay plating curve (open circles) shows a virtually full recovery from the UVB exposure (data from Maher et al. 1979). (b) The influence of a long repair time on the induction of mutations in stationary human fibroblast cells following UVB exposure. The immediate plating mutation curve (open diamonds) is fitted with a purely quadratic equation. The long delay plating curve (open circles) shows a virtually full recovery from the UVB exposure (data from Maher et al. 1979).

(Figure 10.10b) of stationary human fibroblasts for immediate and 7-days delayed plating. After 7 days of repair, the UVB has virtually no effect on either survival or mutation induction.

In addition to these pertinent differences, it is equally important to realise that after UVB or UVC exposure the cells suffer a different 'death' than after ionising radiation exposure. After ionising radiation, cells suffer a mitotic cell death and cells will pass through one or two cell divisions before they cease to divide. After UVB or UVC exposure, the cells are blocked in the DNA synthesis phase and remain there, in a sort of interphase death.

It would be interesting to know whether this block of DNA replication leaves the cells in a sort of senescence, especially in view of the recent articles by Serrano (2017) and Scudellari (2017) in *Nature* on the elimination of senescent cells and the increase in healthy lifespan.

10.5   CONCLUSIONS

The biological action of UVA, which can induce single strand DNA damage, is described by making an analogy with the quadratic (β) coefficient derived for ionising radiation. Consequently, the UVA exposure effect relationships are shown to be purely quadratic.

UVB and UVC, which induce pyrimidine dimers, act in a different way, even though the exposure effects are also purely quadratic. The different approach involves a 'pair' of dimers giving a long-term block to replication which causes interphase death. However, the individual dimers can be repaired during and after exposure until the 'pair' is recognised at DNA replication.

The radiation model developed in Chapters 1 to 9, conveniently adapts to the analysis and understanding of UV exposure.

Part III

Genotoxicology

Part III

11 An Assessment of the Risk of Chemical Mutagens

The model is extended to a consideration of the risk of chemical mutagens by combining the radiation-induced molecular lesion, the DNA double strand break, with the paired double strand lesion block on DNA synthesis induced by UVB and UVC light. A distinction is made between chemical mutagens of which a single molecule can interact with both strands of the DNA helix, which are classified as '1st order' agents, and chemical mutagens of which a single molecule can only interact with one strand of the DNA helix, which are classified as '2nd order' agents. Linear–quadratic exposure response equations are developed for the analysis of the cytotoxic action and mutation induction of cross linking, bi-functional and mono-functional chemical mutagens and are confirmed by the analysis of experimental data. A classification of chemical mutagens into four different groups is convenient to assess the potential risks from the different groups. It is concluded that the Class I group of bi-functional chemical mutagens carry a greater risk than the Class II group of mono-functional chemical mutagens but that any exposure of man to any mutagenic chemical potentially carries a cancer risk and should be subject to strict control and regulation.

11.1 SYNERGISM BETWEEN RADIATION AND CHEMICAL AGENTS

In the late 1970s and early 1980s, a considerable research effort was made to investigate synergism arising from the combination of various chemical agents with radiation. In this case, synergism means that, when two agents have similar effects, then the combination of the two agents gives a greater effect than the sum of the two agents given separately. The aim of the research was to improve the radiotherapy treatment of cancer. Many of the chemical agents used were known to react with DNA in the cell and this led to a mathematical extension of the quantitative radiation model discussed previously in Chapters 1 to 9 (Chadwick et al. 1976; Chadwick and Leenhouts 1981; Leenhouts and Chadwick 1978; Leenhouts et al. 1980). The extension is characterised schematically in Figure 11.1.

This synergistic interaction can be quantified starting from the model for ionising radiation where the number of DNA double strand breaks (N) after a dose (D) is

$$N = \alpha D + \beta D^2. \tag{11.1}$$

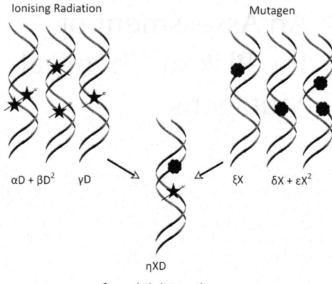

Ionising Radiation   Mutagen

$\alpha D + \beta D^2$ γD ξX $\delta X + \varepsilon X^2$

$\eta X D$

Synergistic Interaction

FIGURE 11.1   A schematic representation of the extension of the ionising radiation model to include the synergistic interaction with a mutagenic agent showing, in particular, the interaction term.

In addition to these double strand breaks, the ionising radiation will also induce a considerable number of DNA single strand breaks (N_s) where

$$N_s = \gamma D. \tag{11.2}$$

By analogy, a mutagenic agent has, in general, the possibility to induce the number of DNA double strand lesions (N_L) after an exposure (X) where

$$N_L = \delta X + \varepsilon X^2 \tag{11.3}$$

and, in addition, a number of DNA single strand lesions (N_{SL}) where

$$N_{SL} = \xi X. \tag{11.4}$$

When the irradiation is combined with an exposure to a mutagen, there is a possibility that the radiation-induced DNA single strand breaks interact with the mutagen-induced single strand lesions to create an additional number of double strand lesions (N_c), where

$$N_c = \eta X D. \tag{11.5}$$

Thus, when radiation is applied after a pre-treatment of (X) with a mutagen, relative survival is given by

$$S_X = \exp\left(-p\left((\alpha + \eta X)D + \beta D^2\right)\right). \tag{11.6}$$

This equation implies that cell survival will be linear–quadratic with dose, the interaction term, which gives the synergy, increases the linear (pα) coefficient with a term that is proportional with the exposure (X) of the pre-treatment and, for a series of radiation survival studies following different levels of pre-treatment, all the survival curves will have the same quadratic (pβ) coefficient.

It is worth noting that not all chemical mutagens will exhibit a linear (δX) coefficient in their effect, as is also the case for UV light. Although the parameter (p), relating the double strand lesions to cell killing, might not be the same for the radiation-induced double strand breaks and mutagen-induced double strand lesions, using different values only complicates the equation. It also appears that it is better to study chemical toxicity by applying different concentrations of the chemical for a constant exposure time, rather than a constant concentration for different exposure times.

An example of the application of the Equation 11.6 to the combination of radiation and a nitrosourea compound is shown in Figure 11.2 (Leenhouts et al. 1980) and Figure 11.3 shows the proportional increase in the linear (pα) coefficient with BCNU (1,3-bis(2-chloroethyl)-1-nitrosourea) pre-treatment. Figure 11.4 shows the purely quadratic cell survival curve with BCNU exposure alone.

A complete chapter (Chapter 11) was devoted to synergy between chemical agents, including base analogues, such as, bromouracil (BUdR), and radiation in Chadwick and Leenhouts (1981).

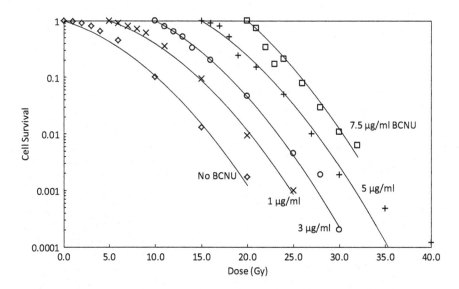

FIGURE 11.2 The radiation survival curves of rat brain tumour cells after pre-treatment with different concentrations of the nitrosourea BCNU. All curves are fitted with the same quadratic (pβ) coefficient (0.011 Gy^{-2}) and the increasing initial linear (pα) coefficient with increasing BCNU treatment is clear to see (data from Wheeler et al. 1977; Deen and Williams 1979). The start of the different survival curves has been staggered by 5 Gy for clarity.

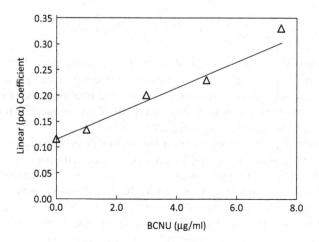

FIGURE 11.3 The increase in the value of the linear (pα) coefficient of the radiation survival curves as a function of the BCNU pre-treatment that was predicted in Equations 11.5 and 11.6. All the survival curves maintained a constant quadratic (pβ) coefficient.

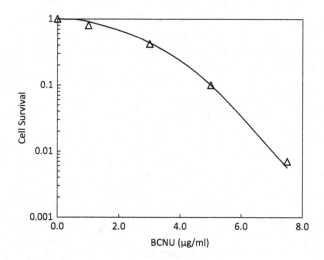

FIGURE 11.4 The cell-killing effect of BCNU, as a function of BCNU concentration, fitted by a purely quadratic function implying that a single molecule of BCNU cannot interact with both strands of the DNA double helix to induce double strand lesions, that is ($\delta = 0$) in Equation 11.3.

11.2 THE EXTENSION OF THE RADIATION MODEL TO QUANTIFY CHEMICAL MUTAGEN CYTOTOXICITY

The successful application of this synergistic analysis to the combination of radiation with a series of mutagenic agents, including UV light, cis-platinum, nitrosourea compounds, halogenated pyrimidine analogues, diamide and misonidazole, led to the development of the radiation model to investigate the toxicity of chemical mutagens (Leenhouts and Chadwick 1984).

The extension of the model to chemical mutagens draws its inspiration from the critical radiation-induced molecular lesion, the DNA double strand break, but also from the double strand deformation block on DNA synthesis, identified as the critical UVB or UVC light-induced lesion. In analogy with the case for ionising radiation and UV light, it is proposed that chemical mutagens can form either one or the other critical double stranded lesion and the mathematical development of analytical functions follows that path. An important distinction between ionising radiation and UV light is the ability of ionising radiation to induce DNA double strand breaks in the passage of a single particle track, in proportion with radiation dose whereas, UV light is unable to do this. Consequently, a distinction is made between chemical mutagens where a single molecule can interact with both strands of the DNA helix, which are classified as '1st order' agents, and chemical mutagens of which a single molecule can only interact with one strand of the DNA helix, which are classified as '2nd order' agents.

A schematic representation of the potential interaction of the two classes of chemical mutagens with DNA in the cell nucleus is presented in Figure 11.5.

11.2.1 THE CALCULATION OF DOUBLE STRAND LESIONS PROPORTIONAL TO EXPOSURE

Some bi- or poly-functional molecules with appropriate dimensions can interact with both strands of the DNA to form, for example, an inter-strand cross-link. The number of these interactions can be calculated as follows:

If (n) is the number of bases per DNA strand per cell and (σ_0) is the probability per base per unit exposure to the chemical mutagen that a double stranded

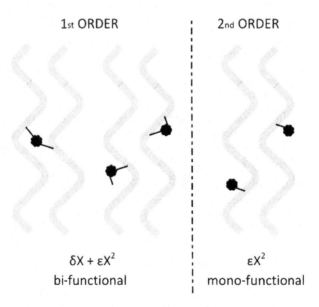

FIGURE 11.5   A schematic representation of the potential interaction of '1st order' and '2nd order' chemical mutagens with the DNA double helix.

lesion is induced, then (N_0) the number of double stranded lesions formed by an exposure (X) is

$$N_0 = n\sigma_0 X. \tag{11.7}$$

If (f_0) is the probability that the double strand lesion is not repaired before it becomes effective, the number of unrepaired double strand lesions is

$$f_0 N_0 = f_0 n\sigma_0 X = \delta X. \tag{11.8}$$

11.2.2 THE CALCULATION OF DOUBLE STRAND LESIONS FROM PAIRS OF SINGLE STRAND LESIONS

A wide variety of chemical mutagens is capable of forming DNA single strand lesions such as intra-strand cross-links, DNA-protein cross-links or alkylations. It is considered unlikely that a single strand lesion will be cytotoxic or mutagenic, unless it 'pairs' with a second single strand lesion to form a paired double strand lesion. The calculation of the number of 'paired' double strand lesions runs as follows.

If (n) is the number of bases per DNA strand per cell then, (2n) is the total number of bases per cell and (σ) is the probability per base per unit exposure that a single strand lesion is formed, then the number of single-strand lesions per cell (π) induced by an exposure (X), is

$$\pi = 2n\sigma X. \tag{11.9}$$

Assuming that a 'paired' DNA double strand lesion is created when a second single strand lesion is formed on the opposite strand from and within (wn) bases of the first single strand lesion then, the average number (N_d) of double strand lesions per cell induced by an exposure (X), is

$$N_d = 2wn^2\sigma^2 X^2. \tag{11.10}$$

Each of the two single strand lesions forming a 'paired' double strand lesion can be repaired independently before the 'paired' double strand lesion is recognised. If (f_1) and (f_2) are the probabilities that the 'first' and 'second' single strand lesions are not repaired before they are recognized as a double strand lesion, then the average number (N) of double strand lesions induced by an exposure (X) and recognised is

$$N = f_1 f_2 N_d = 2f_1 f_2 wn^2\sigma^2 X^2. \tag{11.11}$$

If, after recognition, the double strand lesion can be repaired before it is biologically effective and, if (f_p) is the probability that it is not repaired, then the number of effective double strand lesions is

$$f_p N = 2f_p f_1 f_2 wn^2\sigma^2 X^2 = \varepsilon X^2 \tag{11.12}$$

and the total number of unrepaired, recognised double strand lesions per cell is

$$f_0 N_0 + f_p N_d = f_0 n \sigma_0 X + 2 f_p f_1 f_2 w n^2 \sigma^2 X^2 = \delta X + \varepsilon X^2. \tag{11.13}$$

It is important to recognise that a chemical mutagen might be able to induce more than one type of DNA single strand lesion and that each type of lesion may have its own particular repair process so that the functions (f_1) and (f_2) used here could be compound functions which take account of the different repair processes of the different types of lesion. It is also important to note that it is assumed that the coefficients $(f_0, f_1, f_2$ and $w)$ are independent of the chemical exposure (X) at levels used in the cytotoxic experiments. If the cytotoxic chemical interferes with DNA repair, then the analysis will be complicated.

11.2.3 CYTOTOXICITY

If (p) is the probability that an unrepaired, recognised double strand lesion is cytotoxic, then using Poisson statistics, cell survival (S) is given by

$$S = \exp\left(-p\left(\delta X + \varepsilon X^2\right)\right). \tag{11.14}$$

In analogy with the case of ionising radiation, where one of the induced DNA double strand breaks is considered to be a lethal mutation, in the case of chemical toxicity, just one of the induced double strand lesions should be considered to be the lethal mutation. It is not the cumulative effect of all the induced double strand lesions that causes cell death but the critical disfunction of one.

11.2.4 MUTATION FREQUENCY

Again, in analogy with the case of ionising radiation, the mutation frequency per surviving cell (M_s) will be approximately given by

$$M_s \approx q\left(\delta X + \varepsilon X^2\right). \tag{11.15}$$

11.2.5 EXPOSURE TO A CHEMICAL

The 'dosimetry' of chemical mutagens is complicated by a number of factors which do not confuse the dosimetry of ionising radiation or of UV light. The chemical has to penetrate the cell and the cell nucleus; there are problems of enzymatic activation and enzymatic breakdown, and so on, several of which can be time-dependent. Ehrenberg et al. (1983) have shown that at low levels of exposure the relevant transfer factors can be considered to be first-order, so that 'dose' can be assumed to be a function of the time integral of the concentration level. Thus, if the time of exposure is held constant, then the exposure (X) is directly proportional to the concentration (C) of the chemical. The use of different exposure times can lead to complications and the analysis presented here is not applicable in those cases.

11.3   THE ANALYSIS OF THE CYTOTOXICITY OF CHEMICAL MUTAGENS

A series of analyses of the cytotoxicity of different types of chemical mutagens is presented to illustrate the wide application of Equation 11.14. It is necessary to be aware that the equation may be applied to mutagens which behave in a similar fashion to ionising radiation and are not dependent on the DNA synthesis phase for the recognition of the cytotoxic lesions. Similarly, the equation may be applied to mutagens that do need the DNA synthesis phase to recognise the cytotoxic lesions and which behave like UVB or UVC light.

11.3.1   DNA Cross-Linking Agents

Figures 11.6 to 11.8 present the effect of three different cross-linking agents on cellular toxicity, analysed using Equation 11.14 which is generally applicable for cross-linking agents:

$$S = \exp\left(-p\left(\delta X + \varepsilon X^2\right)\right). \tag{11.14}$$

Figure 11.6 illustrates a cell survival curve of L 1210 mouse cells after a half-hour treatment with nitrogen mustard (Ewig and Kohn 1977) which is proportional with the nitrogen mustard concentration, that is ($\varepsilon = 0$) and

$$S = \exp\left(-p\delta X\right). \tag{11.16}$$

This implies that the DNA inter-strand cross-link is the dominant lesion and that the ($p\delta$) term dominates the cellular survival. The survival curve resembles that which would be expected from a densely ionising radiation exposure.

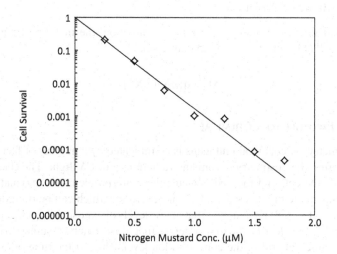

FIGURE 11.6   The survival of mouse cells exposed to different concentrations of nitrogen mustard for 0.5 hr, analysed using Equation 11.14 where ($\varepsilon = 0$). The linear survival curve implies that the inter-strand cross-link lesion is totally dominant (data from Ewig and Kohn 1977).

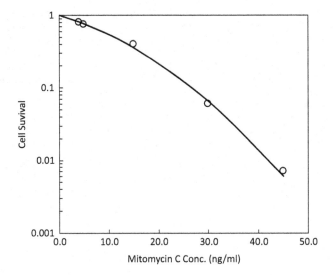

FIGURE 11.7 The survival of mouse cells exposed to different concentrations of mitomycin C, analysed using Equation 11.14 and illustrating a role for both inter-strand cross-links and pairs of single strand lesions in cell killing (data from Hama-Inaba et al. 1983).

Figure 11.7 illustrates a cell survival curve of mouse cells exposed to the cross-linking agent mitomycin C (Hama-Inaba et al. 1983) which exhibits a linear–quadratic function of mitomycin C concentration. This implies that both DNA inter-strand cross-links and pairs of DNA single strand lesions are involved in the lethal action of mitomycin C. The survival curve resembles that which would be expected from a sparsely ionising radiation exposure.

Figure 11.8 presents the survival of mouse cells exposed to 8-methoxypsoralen and near UV light (Arlett et al. 1980). When a cell with 8-methoxypsoralen in the nucleus is exposed to near UV light, a photon of light causes the 8-methoxypsoralen to link to one strand of the DNA, so that DNA inter-strand cross-links arise as a consequence of two UV photons. The survival curve is purely quadratic, that is, ($\delta = 0$) and

$$S = \exp\left(-p\left(\varepsilon X^2\right)\right). \tag{11.17}$$

The survival curve resembles that which would be expected from UV light exposure.

The three different mathematical solutions, derived from the cell-killing effect of DNA cross-linking agents, demonstrate the flexibility of the linear–quadratic Equation 11.14 derived above as the equation for the analysis of the effects of all chemical mutagens. In the following figures, several other types of chemical mutagen will be analysed using the equation although, it has to be said that, in each case, the survival is purely quadratic. This is not surprising as, assuming the processes shown in Figure 11.5 and the mechanism proposed as a potentially lethal lesion, only bi- or poly-functional chemical mutagens with appropriate molecular dimensions would be expected to form DNA inter-strand cross-links in proportion with the

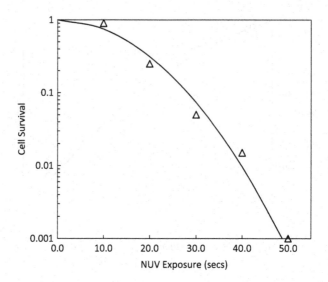

FIGURE 11.8 The survival of mouse cells exposed to near UV light in the presence of 8-methoxypsoralen, analysed using Equation 11.14 where ($\delta = 0$), in accordance with two light photons interacting with the 8-methoxypsoralen to induce one DNA inter-strand cross-link (data from Arlett et al. 1980).

exposure. Mono-functional chemical agents would only be able to induce potentially lethal lesions in proportion with the square of exposure.

This situation is in direct contrast with the situation for ionising radiation where all ionising radiations are able to induce a DNA double strand break in the passage of one particle track. It is worth remembering that, at very low levels of radiation exposure, these DNA double strand breaks, induced in the alpha mode of radiation action, are the ones responsible for the important radiological risk of low-dose radiation. Therefore, it can be concluded that bi- or poly-functional chemical mutagens will probably present a bigger health risk than mono-functional chemical mutagens.

11.3.2 Mono-Functional Chemical Agents

A single molecule of a mono-functional chemical is unable to interact with both DNA strands and the survival of cells exposed to these agents is derived from the general Equation 11.14 as a purely quadratic equation (i.e. $\delta = 0$):

$$S = \exp\left(-p\left(\varepsilon X^2\right)\right). \tag{11.17}$$

Figure 11.9 presents the survival of V79 Chinese hamster cells exposed to the DNA methylating nitrosourea compound MNU (methyl nitrosourea) for 1 hour as a function of MNU concentration (Roberts et al. 1971) fitted with a purely quadratic relationship, in accordance with Equation 11.17.

Figure 11.10 presents the survival of V79 Chinese hamster cells exposed to the DNA methylating compound MMS (methyl methane sulphonate) for 1 hour as a

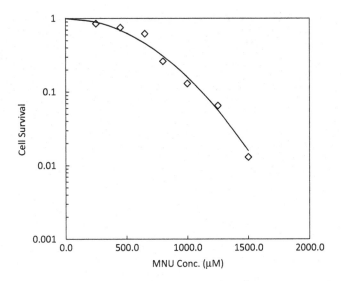

FIGURE 11.9 The survival of V79 Chinese hamster cells exposed to different concentrations of the nitrosourea MNU for 1 hour, analysed using Equation 11.17, so the curved line is a purely quadratic function of MNU concentration (data from Roberts et al. 1971).

function of MMS concentration (Roberts et al. 1971). The line is a purely quadratic relationship, in accordance with Equation 11.17, similar to the case with MNU.

Roberts et al. (1971) also measured the number of DNA methylations induced by MNU, MMS and MNNG (1-methyl-3-nitro-1-nitroso guanidine) which permits an analysis of survival as a function of this molecular damage. The results of this

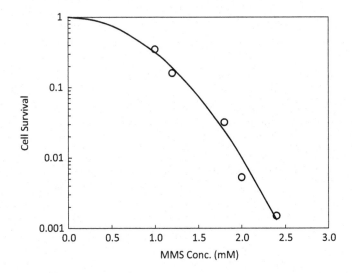

FIGURE 11.10 The survival of V79 Chinese hamster cells exposed to different concentrations of the alkylating agent MMS for 1 hour, analysed using Equation 11.17, so the curved line is a purely quadratic function of MMS concentration (data from Roberts et al. 1971).

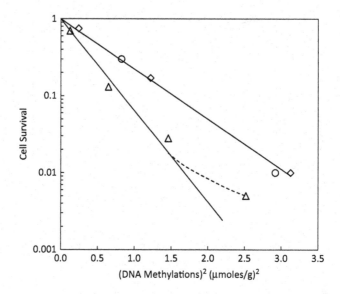

FIGURE 11.11   Cell survival of V79 Chinese hamster cells as a function of the square of the number of DNA methylations induced by MNU (open diamonds) and MMS (open circles) giving the same straight line analysis and MNNG (open triangles) giving a slightly different straight line analysis with a tail-out at lower cell survival (data from Roberts et al. 1971).

analysis are shown in Figure 11.11 where the square of the number of methylations forms the horizontal axis of the graph. This analysis, giving the same straight line correlating the square of the MNU and MMS methylations, and the closeness of the straight line for MNNG, suggests that these chemical mutagens act via the same sort of process and this provides strong support for the molecular mechanisms used in the model which led to Equations 11.14 and 11.17.

The resistant tail on the MNNG data suggests, as was the case for UVB light, that some G_2 cells in the population had more time to repair the DNA methylations prior to recognition in the DNA synthesis phase.

Further evidence in support of the model mechanism can be found in the data of Fraval and Roberts (1979) who measured the excision repair of cis-diamine dichloroplatinum (II)-induced DNA lesions (Pt-DNA) in stationary Chinese hamster cells together with the recovery in cell survival over a period of 3 days. Fraval and Roberts (1979) found an exponential reduction in the number of Pt-DNA lesions with time coupled with a considerable reduction in cell killing over this period. This relationship between improved cell survival and excision repair can be understood using the concepts derived in Chapter 10 for the action of UVB or UVC light based on the proposals of Cleaver and his colleagues (Cleaver 1981; Park and Cleaver 1979a; Cleaver et al. 1979). Park and Cleaver proposed that a pyrimidine dimer could block replication and that pairs of pyrimidine dimers, one on each DNA strand, could form a long-lived block to replication and be a potentially lethal lesion. This is schematically represented for the case of chemical mutagens in Figure 11.12.

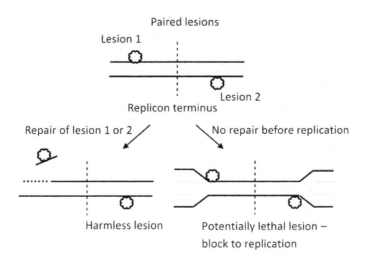

FIGURE 11.12 A schematic representation showing how a pair of chemically induced single strand lesions can form a 'paired' lesion which can block DNA replication and form a potentially lethal (or mutagenic) lesion. If either of the single strand lesions are repaired before replication occurs then replication is not blocked and there is no cellular effect.

This schematic model is completely in line with the quadratic cellular survival curves found for the mono-functional chemical mutagens and Equation 11.17. In its expanded form, this equation carries two parameters, (f_1 and f_2), which are the probabilities that the 'first' and 'second' single strand lesions are not repaired before they are recognised. In the case of mono-functional chemicals, it is very probable that just one type of repair process eliminates the single strand lesions so that ($f_1 = f_2$) and cell survival becomes

$$S = \exp\left(-p\left(2f_p f^2 wn^2 \sigma^2 X^2\right)\right) \tag{11.18}$$

and survival at one exposure level (X_1) with different repair times becomes a function of (f^2), namely:

$$S = \exp\left(-\kappa f^2\right). \tag{11.19}$$

Figure 11.13 presents the correlation between cell survival and the square of the unrepaired Pt-DNA lesions over the repair process, using Equation 11.19.

Figures 11.14 to 11.16 present an analysis of three commonly occurring, environmental chemical mutagens – formaldehyde (Goldmacher and Thilly 1983), ethylene oxide (Poirier and Papadopoulo 1982) and benzo-α-pyrene (Ochi and Oshawa 1983) – using a purely quadratic function and Equation 11.16) The survival after treatment with all three chemical mutagens is directly related to the square of the concentration. The curves, especially in Figure 11.16, illustrate very clearly the extremely low level of cell killing at low concentration levels.

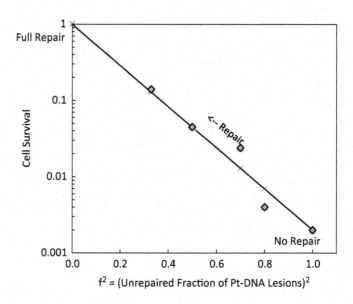

FIGURE 11.13 The correlation between the excision repair of Pt-DNA lesions in stationary Chinese hamster cells over a 3-day period and the improvement in cell survival found by Fraval and Roberts (1979), analysed in accordance with Equation 11.19, which predicts that the logarithm of survival will be proportional to the square of the fraction of unrepaired Pt-DNA lesions.

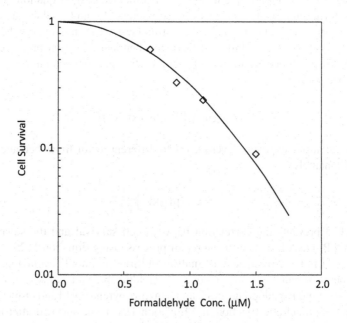

FIGURE 11.14 The survival of human lymphoblast cells exposed for 2 hours to formaldehyde as a function of the exposure concentration, analysed using the purely quadratic Equation 11.17 (data from Goldmacher and Thilly 1983).

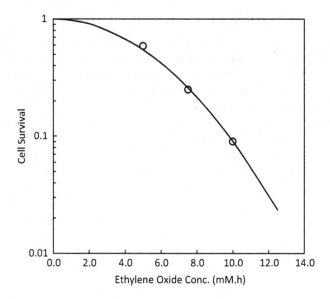

FIGURE 11.15   The survival of human amniotic cells exposed for 1 hour to ethylene oxide as a function of the exposure concentration, analysed using the purely quadratic Equation 11.17 (data from Poirier and Papadopoulo 1982).

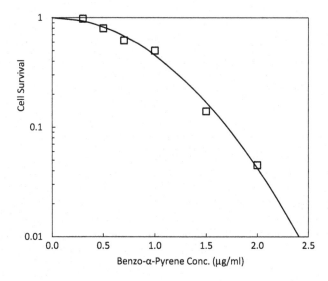

FIGURE 11.16   The survival of Chinese hamster cells exposed for 3 hours to benzo-α-pyrene as a function of the exposure concentration, analysed using the purely quadratic Equation 11.17 (data from Ochi and Oshawa 1983).

11.4   RISK CLASSIFICATION

A classification of chemical mutagens into four different groups is convenient to assess the potential risks from the different groups. A distinction has already been made in Section 11.2, and illustrated in Figure 11.5 between '1st order' and '2nd order' agents. A further differentiation into Class I agents and Class II agents facilitates the assessment of risks and is made on the basis of whether the lesions induced are recognised at the DNA synthesis (S) phase of the cell cycle, like UVB light, or recognised at mitosis, like ionising radiation.

Class I.1 – 1st order – S-phase independent mutagens. These mutagens are bi- or poly-functional chemicals, where a single molecule can interact with both strands of the DNA to induce an inter-strand double strand lesion. It is possible that the molecule can also interact with a single DNA strand to induce a single strand lesion which can be converted into a double strand lesion by a 'second' single strand lesion (see Figure 11.5). These single strand lesions might not all be of the same sort. This does not alter the reasoning and, if the single strand lesions are not the same, either lesion can be the 'first' lesion. This is similar, though not identical, to the case for ionising radiation, although the lesions are not necessarily strand breaks.

The survival curve is given by

$$S = \exp\left(-p\left(\delta X + \varepsilon X^2\right)\right). \tag{11.14}$$

Acute exposure will result in a linear–quadratic cellular effect curve and there will not be a threshold region of no effect. Depending on the repair possibilities of the double strand lesions, it is probable that these agents will have a risk proportional with exposure at low exposure levels, as is the case for ionising radiation. Repair of some of these lesions might occur via micro-homology-enabled recombination repair (MHERR) to create chromosome aberrations if the lesions develop into double strand breaks. Whether or not these agents maintain a quadratic effect after a protracted exposure of several hours or longer, will depend on the speed of repair of the 'first' single strand lesions. This may not be as rapid as is the case for ionising radiation when an interval of a few hours is sufficient for the repair of the 'first' single strand breaks.

Class I.2 – 1st order – S-phase dependent mutagens. These mutagens are bi- or poly-functional chemicals, where a single molecule can interact with both strands of the DNA to induce an inter-strand double strand lesion. It is possible that the molecule can also interact with a single DNA strand to induce a single strand lesion which can pair with another single strand lesion to form a double strand lesion which is recognised when the cell replicates.

The survival curve is given by

$$S = \exp\left(-p\left(\delta X + \varepsilon X^2\right)\right). \tag{11.14}$$

Acute exposure will result in a linear–quadratic cellular effect curve and there will not be a threshold region of no effect. Cells may cease to divide in the DNA replication S-phase, as is the case after UV exposure. Depending on the repair pos-sibilities of the double strand lesions, it is probable that these agents will have a risk proportional with exposure at low exposure levels. Repair of some of these lesions might occur via MHERR to create chromosome aberrations. In a rapidly dividing cell population, the pairs of single strand lesions will be recognised at DNA syn-thesis and the effect curve will maintain a quadratic form. In a stationary, or slowly dividing, cell population, however, the repair of the single strand lesions can proceed independently, eventually to completion, and the effect curve will only be propor-tional with exposure.

Class II.1 – 2nd order – S-phase independent mutagens. These mutagens are mono-functional chemicals, where a single molecule can only interact with one strand of DNA to induce a single strand lesion which can be converted into a double strand lesion by a 'second' single strand lesion on the other strand of the DNA (see Figure 11.5).

The survival curve is given by

$$S = \exp\left(-p\left(\varepsilon X^2\right)\right).$$
(11.17)

Acute exposure will result in a purely quadratic cellular effect curve as a result of the double strand lesions and, although there will not be a threshold region of no effect, the effect at very low exposures will be very small. If the repair of the single strand lesions is relatively rapid and perfect then the 'first' single strand lesions can be repaired before conversion to a double strand lesion. The quadratic (ε) coefficient will decrease as the exposure becomes more chronic and eventually at very long exposure times there might be no measurable effect. The repair of the double strand lesions, which might occur via MHERR to create chromosome aberrations if the lesions become double strand breaks, is unlikely to always be perfect and these will be responsible for the cellular effect.

Class II.2 – 2nd order – S-phase dependent mutagens. These mutagens are mono-functional chemicals, where a single molecule can only interact with one strand of the DNA to induce a single strand lesion. The single strand lesions remain as inde-pendent lesions, and pairs of single strand lesions on opposite DNA strands are only recognised as a 'paired' double strand lesion when the DNA is replicated and the replication process is blocked. Cells will cease to divide in the DNA replication S-phase and suffer an interphase death. This is similar to the case of UVB light (see Chapter 10) although the lesions are not necessarily pyrimidine dimers.

The survival curve is given by

$$S = \exp\left(-p\left(\varepsilon X^2\right)\right).$$
(11.17)

Acute exposure will result in a purely quadratic cellular effect curve and, although there will not be a threshold region of no effect, the effect at very low exposures will be very small. If the repair of the single strand lesions is relatively rapid and perfect then the single strand lesions can be repaired before DNA replication. The quadratic (ε) coefficient will decrease as the time between exposure and DNA replication becomes longer. In stationary cells with long times between exposure and DNA replication, the repair of the single strand lesions will be complete and no cellular effect will be measurable (see Figure 10.6a). The repair of the double strand lesions at DNA replication, which might occur via MHERR to create chromosome aberrations if the lesions become double strand breaks, is unlikely to be perfect and these unrepaired double strand lesions will be responsible for the effect.

It is important to note that the double strand lesions arising from a block of the DNA replication are probably quite different from the double strand lesions arising from the conversion of a 'first' single strand lesion by a 'second' single strand lesion and potentially different repair processes may operate.

The classification of the different chemical mutagens into four different groups has been made solely on the molecular processes involved. These processes lead to the mathematical formulation of cellular effect curves because the quantification of cellular effects leads to a rational consideration of risk.

The *Class I* group of chemical mutagens, whether S-phase independent or not, all have the ability to induce a double strand lesion by the action of a single molecule and will always have an effect proportional with exposure (the linear (δ) coefficient) even for very chronic exposure levels. The repair of these double strand lesions could vary in different cases but the repair is unlikely to be always perfect. These chemicals carry the greatest cytotoxic and mutagenic hazard and, consequently, the greatest cancer risk.

The *Class II* group of chemical mutagens, whether S-phase independent or not, only induce double strand lesions from two single strand lesions and as long as the repair of the single strand lesions is relatively rapid and perfect, chronic exposures to these chemicals will have little or no effect. The effect of chemicals in Class II.2, which depends on the time between exposure and DNA replication, will also show no effect in stationary cells if there is a long period of time between exposure and DNA replication, even after an acute exposure. These Class II chemicals carry much less of a risk from chronic exposure than Class I chemicals, although this should not be seen as an excuse for the widespread use of these Class II mutagens.

The exposure of man to any mutagenic chemical potentially carries a cancer risk and should, therefore, be subject to strict control and regulation.

11.5 CONCLUSIONS

The radiation model developed in Chapters 1 to 9 can be adapted, together with the understanding of the effects of UV exposure, to provide analytical techniques which are useful for different classes of chemical mutagens.

These analytical techniques permit a classification of the risks of different chemical mutagens and could lead to a more generalised approach to the regulation of exposure to both radiation and chemical mutagens.

Epilogue

It is extremely unlikely, at my age, that I will write another textbook on the action of radiation on living matter, so my hope is that this book will stimulate a renewed interest in the role of DNA double strand breaks and their repair in cellular effects, such as cell killing, mutation induction and the formation of chromosomal aberrations as well as in radiation-induced cancer. The model presented in the book really works and offers a clear understanding of radiation biology even though it has not yet acquired widespread use.

I have tried to cover, as comprehensively as possible, the development of the theory from its beginning in 1971 and to present the experimental results of others that support the reasoning and the proposals which make up the theory. The main body of the book, Part I, deals with ionising radiation effects and concludes with some disturbing information on the risks of low-dose radiation. Part II develops an understanding of the effects of UVA light by making an analogy with the quadratic coefficient of the radiation model. It also deals with a different understanding of the effects of UVB and UVC light which was derived as an extension of the radiation model. The combination of this extension with the radiation model led to Part III on the cellular effects and risks of chemical mutagens.

The analysis of the effects of UVA light in Chapter 10, based on the formation of DNA double strand breaks from two independently induced single strand breaks, was developed during the writing of the book thanks to the publication by Cadet et al. (2015). A quantitative measurement of the yield of chromosome aberrations as a function of UVA exposure would, if purely quadratic, provide compelling evidence in favour of the proposal in Chapter 4 that chromosomal aberrations do, indeed, arise from single DNA double strand breaks.

If the content of Chapter 9 on radiological protection seems to be critical of the ICRP, I wish to make clear that this is not my intention. I have a great deal of respect for the aims and achievements of the ICRP. However, the science shows that the 'linear no-threshold' concept, which lies at the base of the current ICRP low-dose radiological risk, does not give any useful information about this risk. The ICRP conclusion that low-dose rate risk is linear with dose from the origin is, to my mind, still valid but the slope of that line will not be revealed in any analysis of cancer incidence in the acutely exposed atomic bomb survivors. The ICRP needs to rebrand with a new approach soon. The same reasoning also applies to the application of the 'linear no-threshold' concept to the derivation of risks for hereditary effects.

UNSCEAR is another international organisation for which I have the utmost respect as a source of extensive and thorough reviews of the ongoing study of the effects of radiation on the population. A rapid and balanced overview of the field can be gained by consulting the UNSCEAR publications.

Although I have never been involved in radiotherapy, I know that alpha/beta ratios derived from the linear–quadratic dose–effect relationship have been widely used in fractionated regimes to improve the killing of tumour cells while sparing the effect on the surrounding normal healthy tissue cells. I hope that a consideration of the parameterisation of the alpha and beta coefficients, developed in Chapter 1, will lead to further improvements in radiotherapy because these coefficients are affected differently by changes in exposure conditions. For instance, the combination of the linear–quadratic dose–effect relationship with the DNA double strand break leads to the prediction of synergy with chemical agents that attack DNA (see Section 11.1). This leads to the expectation on theoretical grounds that a base analogue, such as bromouracil (BUdR), should be useful in therapy as a sensitiser of sparsely ionising radiation in tumour cells compared with normal healthy tissue. The rapidly dividing tumour cells should accumulate much more of the base analogue during DNA replication than slowly dividing normal tissue cells and the base analogue sensitises the radiation effect in the tumour cells by considerably increasing the linear (α) coefficient (see Figure 3.10). I cannot understand, therefore, why base analogues are not used more widely in radiation therapy. The synergistic interaction of base analogues with radiation is discussed in some detail in Section 11.5.2 of Chadwick and Leenhouts (1981). Independent of the role of synergy, the alpha coefficient is normally dependent on radiation quality whereas the beta coefficient is affected by dose rate but is also more strongly influenced by changes in the cellular environment during exposure. The alpha/beta ratio is unlikely to be constant and the current trend to hypo-fractionation therapy regimes, with larger treatment doses and fewer treatment sessions, means that the beta coefficient will play a more important role in the cell-killing effect.

Some readers will, I am sure, think that much of the data used in the book are taken from rather old publications. However, the quality of the science is not reduced by age and, for a retired scientist with no institutional connection, it can prove to be an expensive business trying to get hold of relatively modern data from recent publications, even with 'Open Access'.

Some readers may also think that I have ignored some recent developments in ionising radiation biology, such as radiation-induced 'bystander effects' and radiation-induced 'genomic instability', but I have serious misgivings about the validity of much of this work, as evidenced by my publication in 2016 (Chadwick, 2016).

The work contained in the book covers 45 years of research moving from early insights, interesting developments, supportive data from others, strong criticisms (which continue), enlightenment, building on the work of others (chromosome aberrations and UV light), and extension to other DNA damaging agents, to gain a specific overview of the field. During these 45 years, I have waited for the theory to be disproved convincingly. Fortunately for me, that proof has not been forthcoming and so I commend the theory to you.

My abiding regret during the preparation of the book is that I was unable to persuade my colleague, Henk Leenhouts, to come out of retirement and join the project. We have enjoyed some 45 years of friendship and collaborative research and without him the concepts, analyses and insights gained into radiation biology presented in this book would never have been developed.

I have been planning to write this text book for several years but never quite got round to it and, instead, I prepared an easy-going, popular science version of the model ('Rays' Awareness: Radiation Health Effects Made Easy with Professor Dee and Doctor Hay', www.yps-publishing.co.uk). Once that was finished, I finally felt in the right frame of mind to start this text book. It has not been easy but, it has been very satisfying. I have especially to pay tribute to, and thank, my wife, Hilary, for her continuing encouragement, infinite patience and the invaluable correction of my long-winded English prose, making sense out of nonsense. Our lovely Bichon Frisé, Bonnie, has been a constant companion and has kept me grounded by reminding me when it was time for her daily grooming session … 'Woof, time to close the computer, Dad!'

References

Arlett CF, Heddle JA, Broughton BC, Rogers AM (1980) Cell killing and mutagenesis by 8-methoxypsoralen in mammalian (rodent) cells. *Clin. Exp. Dermatol.* **5**(2): 147–158.

Armitage P, Doll R (1954) The age distribution of cancer and a multi-stage theory of carcinogenesis. *Br. J. Cancer* **8**(1): 1–12.

Arnoult N, Correai A, Ma J, et al. (2017) Regulation of DNA repair pathway choice in S and G2 phases by the NHEJ inhibitor CYREN. *Nature* **549**(7673): 548–552.

Astrahan M (2008) Some implications of linear-quadratic-linear radiation dose-response with regard to hypofractionation. *Med. Phys.* **35**(9): 4161–4172.

Bailey SM, Cornforth MN (2007) Telomeres and DNA double strand breaks: Ever the twin shall meet. *Cell. Mol. Life Sci.* **64**(22): 2956–2964.

Bailey SM, Cornforth MN, Ullrich RL, Goodwin EH (2004) Dysfunctional mammalian telomeres join with DNA double-strand breaks. *DNA Repair* **3**(4): 349–357.

Bakhoum SF, Ngo B, Laughney AM, et al. (2018) Chromosomal instability drives metastasis through a cytosolic DNA response. *Nature* **553**(7689): 467–472.

Bandopadhayay P, Meyerson M (2018) Landscapes of childhood tumours. *Nature* **555**(7696): 316–317.

Barendsen GW (1964) Impairment of the proliferative capacity of human cells in culture by α-particles with differing linear energy transfer. *Int. J. Radiat. Biol. Relat. Stud. Phys. Chem. Med.* **8**(5): 453–466.

Barendsen GW, Koot CJ, Van Kersen GR, Bewley DK, Field SB, Parnell CJ (1966) The effect of oxygen on the impairment of the proliferative capacity of human cells in culture by ionizing radiations of different LET. *Int. J. Radiat. Biol. Relat. Stud. Phys. Chem. Med.* **10**(4): 317–327.

Becker K (1997) Threshold or no threshold, that is the question. *Radiat. Prot. Dosim.* **71**(1): 3–5.

Bedford JS, Cornforth MN (1987) Relationship between the recovery from sub-lethal X-ray damage and the rejoining of chromosome breaks in normal human fibroblasts. *Radiat. Res.* **111**(3): 406–423.

Benbow RM, Gaudette MF, Hines PJ, Shioda M (1985) Initiation of DNA replication in eukaryotes. In: Boynton AL, Leffert HL (eds.) *Control of Animal Cell Proliferation*, Vol 1 (Academic Press, New York) pp 449–483.

Benham CJ (1979) Torsional stress and local denaturation in supercoiled DNA. *Proc. Natl. Acad. Sci. USA*. **76**(8): 3870–3874.

Besaratinia A, Kim SI, Pfeifer GP (2008) Rapid repair of UVA-induced oxidized purines and persistence of UVB-induced dipyrimidine lesions determine the mutagenicity of sunlight in mouse cells. *FASEB J.* **22**(7): 2379–2392.

Bhambhani R, Kuspira J, Giblak RE (1973) A comparison of cell survival and chromosomal damage using CHO cells synchronized with and without colcemid. *Can. J. Genet. Cytol.* **15**(3): 605–618.

Bird R, Burki J (1971) Inactivation of mammalian cells at different stages of the cell cycle as a function of radiation linear energy transfer. In: *Biophysical Aspects of Radiation Quality. Technical Report Series* (International Atomic Energy Agency, Vienna) pp 241–249.

Bodgi L, Canet A, Pujo-Menjouet L, Lesne A, Victor JM, Foray N (2016) Mathematical models of radiation action on living cells: From the target theory to the modern approaches. A historical and critical review. *J. Theor. Biol.* **394**: 93–101.

Bodgi L, Foray N (2016) The nucleo-shuttling of the ATM protein as a basis for a novel theory of radiation response: Resolution of the linear-quadratic model. *Int. J. Radiat. Biol.* **92**(3): 117–131.

Bond VP, Wielopolski L, Shani G (1996) Current misinterpretations of the linear no-threshold hypothesis. *Health Phys.* **70**(6): 877–882.

Bouffler SD, Breckon G, Cox R (1996) Chromosomal mechanisms in murine radiation acute myeloid leukaemogenesis. *Carcinogenesis* **17**(4): 655–659.

Brenner DJ (2008) The linear-quadratic model is an appropriate methodology for determining isoeffective doses at large doses per fraction. *Semin. Radiat. Oncol.* **18**(4): 234–239.

Brianese RC, Nakamura KDM, Almeida FGDSR, et al. (2018) BRCA1 deficiency is a recurrent event in early-onset triple-negative breast cancer: A comprehensive analysis of germline mutations and somatic promoter methylation. *Breast Cancer Treat. Res.* **167**(3): 803–814.

Bridges BA (2008) Effectiveness of tritium beta particles. *J. Radiol. Prot.* **28**(1): 1–3.

Britten RJ, Kohne DE (1968) Repeated sequences in DNA. *Science* **161**(3841): 529–540.

Brown R, Thacker J (1980) Characterization of radiation-induced mutants of cultured mammalian cells. *Radiat. Environ. Biophys.* **17**(4): 341.

Buckel TB, Goldstein AM, Fraser MC, Rogers B, Tucker MA (2006) Recent tanning bed use: A risk factor for melanoma. *Arch. Dermatol.* **142**(4): 485–488.

Burch PRJ (1957) Calculations of energy dissipation characteristics in water for various radiaions. *Radiat. Res.* **6**(3): 289–301.

Burch PRJ (1960) Radiation carcinogenesis: A new hypothesis. *Nature* **185**: 135–142.

Burke EA, Petit RM (1960) Absorption analysis of X-ray spectra produced by beryllium window tubes operated at 20 and 50 kVp. *Radiat. Res.* **13**(2): 271–285.

Cadet J, Douki T, Ravanat J-L (2015) Oxidatively generated damage to cellular DNA by UVB and UVA radiation. *Photochem. Photobiol.* **91**(1): 140–155.

Calabrese EJ (2017) Flaws in the LNT single-hit model for cancer risk: An historical assessment. *Environ. Res.* **158**: 773–788.

Cancer Genome Atlas Research Network (2017a) Integrated genomic characterization of oesophageal carcinoma. *Nature* **541**(7636): 169–175.

Cancer Genome Atlas Research Network (2017b) Integrated genomic and molecular characterization of cervical cancer. *Nature* **543**(7645): 378–384.

Cejka P (2017) Biochemistry: Complex assistance for DNA invasion. *Nature* **550**(7676): 342–343.

Chadwick KH (2016) Non-targeted effects and radiation-induced cancer. *J. Radiol. Prot.* **36**(4): 1011–1014.

Chadwick KH (2017) Towards a new dose and dose-rate effectiveness factor (DDREF)? Some comments. *J. Radiol. Prot.* **37**(2): 422–433.

Chadwick KH, Leenhouts HP (1973a) A molecular theory of cell survival. *Phys. Med. Biol.* **18**(1): 78–87.

Chadwick KH, Leenhouts HP (1973b) The molecular target theory of cell survival and its application in radiobiology. *Euratom Report* EUR 4918 (CEC, Luxembourg).

Chadwick KH, Leenhouts HP (1974) Chromosome aberrations and cell death. In: Booz J, Ebert HG, Eickel R, Waker A (eds.) *Fourth Symposium on Microdosimetry.* EUR 5122 (CEC, Luxembourg) pp 585–605.

Chadwick KH, Leenhouts HP (1975) The effect of an asynchronous population of cells on the initial slope of dose-effect curves. In: Alper T (ed.) *Cell Survival after Low Doses of Radiation: Theoretical and Clinical Implications* (Institute of Physics and John Wiley & Sons, London) pp 57–63.

Chadwick KH, Leenhouts HP (1978) The rejoining of DNA double-strand breaks and a model for the formation of chromosomal rearrangements. *Int. J. Radiat. Biol. Relat. Stud. Phys. Chem. Med.* **33**(6): 517–529.

Chadwick KH, Leenhouts HP (1981) *The Molecular Theory of Radiation Biology* (Springer Verlag, Heidelberg).

Chadwick KH, Leenhouts HP (1983) A model for the cytotoxic action of UV. *Phys. Med. Biol.* **28**(12): 1369–1383.

Chadwick KH, Leenhouts HP (1995) Multi-step carcinogenesis and the implications for low dose radiation philosophy. *Trans. Inst. Chem. Eng. B* **73B**: S18–S23.

Chadwick KH, Leenhouts HP (2011a) Radiation-induced cancer arises from a somatic mutation. *J. Radiol. Prot.* **31**(1): 41–48.

Chadwick KH, Leenhouts HP (2011b) Reply to comment on 'Radiation-induced cancer arises from a somatic mutation'. *J. Radiol. Prot.* **31**(3): 372.

Chadwick KH, Leenhouts HP, Brugmans MJP (2002) Implications of the analysis of epidemiological data using a two-mutation carcinogenesis model for radiation risks. *Radiat. Prot. Dosim.* **99**(1–4): 265–268.

Chadwick KH, Leenhouts HP, Laheij GMH, Venema LB (1995) The implications of a two-mutation carcinogenesis model for internal emitters. In: van Kaick G, Karaoglou A, Kellerer AM (eds.) *Health Effects of Internally Deposited Radionuclides: Emphasis on Radium and Thorium* (World Scientific, Singapore) pp 353–360.

Chadwick KH, Leenhouts HP, Sijsma MJ (1998) A comprehensive analysis of dose fractionation [unpublished].

Chadwick KH, Leenhouts HP, Szumiel I, Nias AHW (1976) An analysis of the interaction of a platinum complex and radiation with CHO cells using the molecular theory of cell survival. *Int. J. Radiat. Biol. Relat. Stud. Phys. Chem. Med.* **30**(6): 511–524.

Chapman JD, Dugle DL, Reuvers AP, Gillespie CJ, Borsa J (1975b) Chemical radio-sensitization studies with mammalian cells growing *in vitro*. In: Nygaard OF, Adler HI, Sinclair WK (eds.) *Radiation Research-Biomedical, Chemical and Physical Perspectives* (Academic Press, New York) pp 752–760.

Chapman JD, Gillespie CJ, Reuvers AP, Dugle DL (1975a) The inactivation of Chinese hamster cells by X-rays: The effects of chemical modifiers on single- and double-events. *Radiat. Res.* **64**(2): 365–375.

Chapman JD, Reuvers AP, Doern SD, Gillespie CJ, Dugle DL (1976) Radiation chemical probes in the study of mammalian cell inactivation and their influence on radiobiological effectiveness. In: Booz J, Ebert HG, Smith BGR (eds.) *Fifth Symposium on Microdosimetry*. EUR 5452 (CEC, Luxembourg) pp 775–793.

Christensen RC, Tobias CA, Taylor WD (1972) Heavy-ion induced single- and double- strand breaks in *ΦX-174* replicative form DNA. *Int. J. Radiat. Biol. Relat. Stud. Phys. Chem. Med.* **22**(5): 457–477.

Cleaver JE (1981) Inhibition of DNA replication by hydroxyurea and caffeine in an ultraviolet-irradiated human fibroblast cell line. *Mutat. Res.* **82**(1): 159–171.

Cleaver JE, Thomas GH, Park SD (1979) *Xeroderma pigmentosum* variants have slow recovery of DNA synthesis after irradiation with ultraviolet light. *Biochem. Biophys. Acta* **564**(1): 122–131.

Coggle JE (1988) Lung tumour induction in mice after X-rays and neutrons. *Int. J. Radiat. Biol. Relat. Stud. Phys. Chem. Med.* **53**(4): 585–598.

Cooke BC, Fielden EM, Johnson M (1976) Polyfunctional radiosensitizers. I. Effects of a nitroxyl biradical on the survival of mammalian cells *in vitro*. *Radiat. Res.* **65**(1): 152–162.

Cook WD, McCaw BJ, Herring C, et al. (2004) PU.1 is a suppressor of myeloid leukemia, inactivated in mice by gene deletion and mutation of its DNA binding domain. *Blood* **104**(12): 3437–3444.

Cormack DV, Johns HE (1952) Electron energies and ion densities in water irradiated with 200 keV, 1 MeV, and 25 MeV radiation. *Br. J. Radiol.* **24**(295): 369–381.

Cornforth MN (1990) Testing the notion of the one-hit exchange. *Radiat. Res.* **121**(1): 21–27.

Cornforth MN, Anur P, Wang N, et al. (2018) Molecular cytogenetics guides massively parallel sequencing of radiation-induced chromosome translocation in human cells. *Radiat. Res.* **190**(1): 88–97.

Cornforth MN, Bedford JS (1983) X-ray-induced breakage and rejoining of human interphase chromosomes. *Science* **222**(4628): 1141–1143.

Cornforth MN, Bedford JS (1987) A quantitative comparison of PLD repair and the rejoining of interphase chromosome breaks in low passage normal human diploid fibroblasts. *Radiat. Res.* **111**(3): 385–405.

Corry PM, Cole A (1973) Double-strand rejoining in mammalian DNA. *Nat. New Biol.* **245**(143): 100–101.

Cox R, Mason WK (1978) Do radiation-induced thioguanine resistant mutants of cultured mammalian cells arise by HGPRT gene mutation or X-chromosome rearrangement? *Nature* **276**(5688): 629–630.

Cox R, Masson WK (1979) Mutation and inactivation of cultured mammalian cells exposed to beams of accelerated heavy ions. III. Human diploid fibroblasts. *Int. J. Radiat. Biol. Relat. Stud. Phys. Chem. Med.* **36**(2): 149–160.

Cox R, Thacker J, Goodhead DT (1977) Inactivation and mutation of cultured mammalian cells by aluminium characteristic ultrasoft X-rays. II. Dose-responses of Chinese hamster and human diploid cells to aluminium X-rays and radiations of different LET. *Int. J. Radiat. Biol. Relat. Stud. Phys. Chem. Med.* **31**(6): 561–576.

Cox R, Thacker J, Goodhead DT, Munson RJ (1977) Mutation and inactivation of cultured mammalian cells by various ionizing radiations. *Nature* **267**: 425–427.

Davidson EH, Britten RJ (1973) Organization, transcription, and regulation in the animal genome. *Q. Rev. Biol.* **48**(4): 565–613.

Davidson EH, Hough BR, Amemson CS, Britten RJ (1973) General interspersion of repetitive with non-repetitive sequence elements in the DNA of *Xenopus*. *J. Mol. Biol.* **77**(1): 1–23.

Deen DF, Williams ME (1979) Isobologram analysis of X-ray-BCNU interactions *in vitro*. *Radiat. Res.* **79**(3): 483–491.

de Sousa e Melo F, Kurtova AV, Harnoss JM, et al. (2017) A distinct role for Lgr5+ stem cells in primary and metastatic colon cancer. *Nature* **543**(7647): 676–680.

Dewey WC, Furman SC, Miller HH (1970) Comparison of lethality and chromosomal damage induced by X-rays in synchronized Chinese hamster cells *in vitro*. *Radiat. Res.* **43**(3): 561–581.

Dewey WC, Stone LE, Miller HH, Giblak RE (1971a) Radiosensitisation with 5-bromodeoxyuridine of Chinese hamster cells irradiated during different phases of the cell cycle. *Radiat. Res.* **47**(3): 672–688.

Dewey WC, Miller HH, Leeper DB (1971b) Chromosomal aberrations and mortality of X-irradiated mammalian cells: Emphasis on repair. *Proc. Natl. Acad. Sci. USA.* **68**(3): 667–671.

Dewey WC, Sapareto SA, Betten DA (1978) Hyperthermic radiosenitization of synchronous Chinese hamster cells: relationship between lethality and chromosomal aberrations. *Radiat. Res.* **76**(1): 48–59.

Dikomey E (1982) Effect of hyperthermia at 42 and 45°C on repair of radiation-induced DNA strand breaks in CHO cells. *Int. J. Radiat. Biol. Relat. Stud. Phys. Chem. Med* **41**(6): 603–614.

Douglas ME, Ali FA, Costa A, Diffley JFX (2018) The mechanism of eukaryotic CMG helicase activation. *Nature* **555**(7695): 265–268.

Douki T, Reynaud-Angelin A, Cadet J, Sage E (2003) Bipyrimidine photoproducts rather than oxidative lesions are the main type of DNA damage involved in the genotoxic effect of solar UVA radiation. *Biochemistry* **42**(30): 9221–9226.

Dugle DL, Gillespie CJ (1975) Kinetics of single-strand repair mechanism in mammalian cells. In: Hanawalt PC, Setlow RB (eds.) *Molecular Mechanisms for Repair of DNA* (Plenum, New York) pp 685–687.

Dugle DL, Gillespie CJ, Chapman JD (1976) DNA strand breaks, repair, and survival in X-irradiated mammalian cells. *Proc. Natl. Acad. Sci. USA.* **73**(3): 809–812.

Dunn JM, Philips RA, Becker AJ, Gallie BL (1988) Identification of germline and somatic mutations affecting the retinoblastoma gene. *Science* **241**(4874): 1797–1800.

Edwards AA, Lloyd DC (1980) On the prediction of dose-rate effects for dicentric production in human lymphocytes by X- and γ-rays. *Int. J. Radiat. Biol. Relat. Stud. Phys. Chem. Med.* **37**(1): 89–92.

Ehrenberg LE, Moustacchi E, Osterman-Golkar S (1983) Dosimetry of genotoxic agents and dose-response relationships of their effects. *Mutat. Res.* **123**(2): 121–182.

Encinas G, Sabelnykova VY, de Lyra EC, et al. (2018) Somatic mutations in early onset luminal breast cancer. *Oncotarget* **9**(32): 22460–22479.

Ewig RAG, Kohn KW (1977) DNA damage and repair in mouse leukemia cells treated with nitrogen mustard (HN2), 1,3-bis (2-chloroethyl)-1-nitrosourea and other nitrosoureas. *Cancer Res.* **37**(7 Pt 1): 2114–2122.

Fairlie I (2007) RBE and w_R values of Auger emitters and low-range beta emitters with particular reference to tritium. *J. Radiol. Prot.* **27**(2): 157–168.

Fang X, Ide N, Higashi SI, et al. (2014) Somatic cell mutations caused by 365 nm LED UVA due to DNA double-strand breaks through oxidative damage. *Photochem. Photobiol. Sci.* **13**(9): 1338–1346.

Fell LJ, Paul ND, McMillan TJ (2002) Role for non-homologous end-joining in the repair of UVA-induced DNA damage. *Int. J. Radiat. Biol.* **78**(11): 1023–1027.

Folkard M, Prise KM, Vojnovic B, Davies S, Michael BD (1989) The irradiation of V79 mammalian cells by protons with energies below 2 MeV. Part I: Experimental arrangement and measurements of cell survival. *Int. J. Radiat. Biol.* **56**(3): 221–237.

Frankenberg D, Kelnhofer K, Bar K, Frankenberg-Schwager M (2002) Enhanced neoplastic transformation by mammography X-rays relative to 200 kVp X-rays: Indication for a strong dependence on photon energy of the RBE_M for various end points. *Radiat. Res.* **157**(1): 99–105.

Franken NA, Oei AL, Kok HP, et al. (2013) Cell survival and radiosensitisation: Modulation of the linear and quadratic parameters of the LQ model (Review). *Int. J. Oncol.* **42**(5): 1501–1515.

Fraval HNA, Roberts JJ (1979) Excision repair of cis-diamine dichloroplatinum (II) – Induced damage to DNA of chinese hamster cells. *Cancer Res.* **39**(5): 1793–1797.

Gardner HA, Gallie BL, Knight LA, Phillips RA (1982) Multiple karyotypic changes in retinoblastoma tumor cells: Presence of normal chromosome No.13 in most tumors. *Cancer Genet. Cytogenet.* **6**(3): 201–211.

Gaudette MF, Benbow RM (1986) Replication forks are underrepresented in chromosomal DNA of *Xenopus laevis* embryos. *Proc. Natl. Acad. Sci. USA.* **83**(16): 5953–5947.

Gillespie CJ, Chapman JD, Reuvers AP, Dugle DL (1975a) The inactivation of Chinese hamster cells by X-rays: Synchronized and exponential cell populations. *Radiat. Res.* **64**(2): 353–364.

Gillespie CJ, Chapman JD, Reuvers AP, Dugle DL (1975b) Survival of X-irradiated hamster cells: Analysis in terms of the Chadwick-Leenhouts model. In: Alper T (ed.) *Cell Survival after Low Doses of Radiation: Theoretical and Clinical Implications* (Institute of Physics & John Wiley & Sons, London) pp 25–31.

Goldmacher VS, Thilly WG (1983) Formaldehyde is mutagenic for cultured human cells. *Mutat. Res.* **116**(3–4): 417–422.

Goodhead DT (1980) Models of radiation inactivation and mutagenesis. In: Meyn RE, Withers HR (eds.) *Radiation Biology in Cancer Research* (Raven Press, New York) pp 231–247.

Goodhead DT (2009) The relevance of dose for low-energy beta emitters. *J. Radiol. Prot.* **29**(3): 321–333.

Goodhead DT, Munson RJ, Thacker J, Cox R (1980) Mutation and inactivation of cultured mammalian cells exposed to beams of accelerated heavy ions. IV. Biophysical interpretation. *Int. J. Radiat. Biol. Relat. Stud. Phys. Chem. Med.* **37**(2): 135–167.

Goodhead DT, Thacker J (1977) Inactivation and mutation of cultured mammalian cells by aluminium characteristic ultrasoft X-rays I. Properties of aluminium X-rays and preliminary experiments with Chinese Hamster cells. *Int. J. Radiat. Biol. Relat. Stud. Phys. Chem. Med.* **31**(6): 541–559.

Goodhead DT, Thacker J, Cox R (1979) Effectiveness of 0.3 keV carbon ultrasoft X-rays for the inactivation and mutation of cultured mammalian cells. *Int. J. Radiat. Biol. Relat. Stud. Phys. Chem. Med.* **36**(2): 101–114.

Goodhead DT, Virsik RP, Harder D, Thacker J, Cox R, Blohm R (1981) Ultrasoft X-rays as a tool to investigate radiation-induced dicentric chromosome aberrations. In: Booz J, Ebert HG, Hartfiel HD (eds.) *Seventh Symposium on Microdosimetry*. EUR 7147 (Harwood Academic, London) pp 1275–1285.

Grade M, Difilippantonio MJ, Camps J (2015) Patterns of chromosomal aberrations in solid tumors. *Cancer Res.* **200**: 115–142.

Grahn D, Lombard LS, Carnes BA (1992) The comparative tumorigenic effects of fission neutrons and cobalt-60 γ rays in the B6CF$_1$ mouse. *Radiat. Res.* **129**(1): 19–36.

Greaves M, Maley CC (2012) Clonal evolution in cancer. *Nature* **48**(7381): 306–313.

Greinert RB, Volkmer S, Henning EW, et al. (2012) UVA-induced DNA double-strand breaks result from the repair of clustered oxidative DNA damages. *Nucleic Acids Res.* **40**(20): 10263–10273.

Greten FR (2017) Cancer: Tumour stem-cell surprises. *Nature* **543**(7647): 626–627.

Griffin CS, Hill MA, Papworth DG, Townsend KMS, Savage JRK, Goodhead DT (1998) Effectiveness of 0.28 keV carbon K X-rays at producing simple and complex chromosome exchanges in human fibroblasts *in vitro* detected using FISH. *Int. J. Radiat. Biol.* **73**(6): 591–598.

Griffin CS, Stevens DL, Savage JRK (1996) Ultrasoft 1.5 keV aluminium X-rays are efficient producers of complex chromosome exchange aberrations revealed by fluorescence *in situ* hybridization. *Radiat. Res.* **146**(2): 144–150.

Gröbner SN, Worst BC, Weischenfeldt J, et al. (2018) The landscape of genomic alterations across childhood cancers. *Nature* **555**(7696): 321–327.

Hahn S, Buratowski S (2016) Snapshots of transcription initiation. *Nature* **533**(7603): 331–332.

Hall EJ, Gross W, Dvorak RF, Kellerer AM, Rossi HH (1972) Survival curves and age response functions for Chinese hamster cell exposed to X-rays or high LET alpha-particles. *Radiat. Res.* **52**(1): 88–98.

Hama-Inaba H, Hieda-Shiomi N, Shiomi T, Sato K (1983) Isolation and characterization of mitomycin C sensitive mouse lymphoma cell mutants. *Mutat. Res.* **108**(1–3): 405–416.

Harrison J (2009) Doses and risks from tritiated water and environmental organically bound tritium. *J. Radiol. Prot.* **29**(3): 335–349.

Herr L, Friedrich T, Durante M, Scholz M (2015) A comparison of kinetic photon cell survival models. *Radiat. Res.* **184**(5): 494–508.

Heyes GJ, Mill AJ (2004) The neoplastic transformation potential of mammography X-rays and atomic bomb spectrum radiation. *Radiat. Res.* **162**(2): 120–127.

Heyes GJ, Mill AJ, Charles MW (2006) Enhanced biological effectiveness of low energy X-rays and implications for the UK breast screening programme. *Br. J. Radiol.* **79**(939): 195–200.

Heyes GJ, Mill AJ, Charles MW (2009) Mammography—Oncogenicity at low doses. *J. Radiol. Prot.* **29**(2A): A123–A132.

Holliday R (1964) A mechanism for gene conversion in fungi. *Genet. Res.* **5**(2):282–304

Hornsey S (1973) The effectiveness of fast neutrons compared with low LET radiation on cell survival in the mouse jejunum. *Radiat. Res.* **55**(1): 58–68.

Howe GR (1995) Lung cancer mortality between 1950 and 1987 after exposure to fractionated moderate-dose-rate ionizing radiation in the Canadian fluoroscopy cohort study and a comparison with lung cancer mortality in the Atomic Bomb Survivors study. *Radiat. Res.* **142**(3): 295–304.

ICRP (1991) 1990 recommendations of the International Commission on Radiological Protection. ICRP Publication No. 60. *Ann. ICRP* 21.

ICRP (2003) Relative biological effectiveness, radiation weighting and quality factor. ICRP Publication No. 92. *Ann. ICRP* 33.

ICRP (2007) The 2007 recommendations of the International Commission on Radiological Protection. ICRP Publication No. 103. *Ann. ICRP* 7: 2–4.

ICRU (1970) Report 16. *Linear Energy Transfer* (International Commission on Radiation Units and Measurements, Washington).

Ikebuchi M, Osmak M, Han A, Hill CK (1988) Multiple, small exposures of far-ultraviolet or mid-ultraviolet light change the sensitivity to acute ultraviolet exposures measured by cell lethality and mutagenesis in V79 Chinese hamster cells. *Radiat. Res.* **114**(2): 248–267.

Iliakis G (1984a) The influence of conditions affecting repair and fixation of potentially lethal damage on the induction of 6-thioguanine resistance after exposure of mammalian cells to X-rays. *Mutat. Res.* **126**(2): 215–225.

Iliakis G (1984b) The influence of PLD repair on the induction frequency of 6-thoiguanine resistant mutants in Ehrlich ascites tumour cells after exposure to X-rays and ultraviolet light. *Br. J. Cancer* **49**: Suppl. VI, 113–117.

Iliakis G, Nusse M (1982) Conditions supporting repair of potentially lethal damage cause significant reduction of ultra-violet light induced division delay in synchronized and plateau-phase Ehrlich ascites tumour cells. *Radiat. Res.* **91**(3): 483–506.

Jenner TJ, deLara CM, O'Neill P, Stevens DL (1993) Induction and rejoining of DNA double strand breaks in V79-4 mammalian cells following γ- and α- irradiation. *Int. J. Radiat. Biol.* **64**(3): 265–273.

Johns HE (1969) X-rays and teleisotope γ-rays. In: Attix FH, Tochilin E (eds.) *Radiation Dosimetry*, Vol III (Academic Press, New York) pp 1–50.

Kavathas P, Bach FH, De Mars R (1980) Gamma ray-induced loss of expression of HLA and glyoxylase I alleles in lymphoblastoid cells. *Proc. Natl. Acad. Sci. USA.* **77**(7): 4251–4255.

Kellerer AM (1975) Statistical and biophysical aspects of the survival curve. In: Alper T (ed.) *Cell Survival after Low Doses of Radiation: Theoretical and Clinical Implications* (Institute of Physics & John Wiley & Sons, London) pp 69–77.

Kellerer AM, Rossi HH (1971) RBE and the primary mechanism of radiation action. *Radiat. Res.* **47**(1): 15–31.

Kellerer AM, Rossi HH (1972) The theory of dual radiation action. *Curr. Top. Radiat. Res. Q.* **8**: 85–158.

Kellerer AM, Rossi HH (1978) A generalized formulation of dual radiation action. *Radiat. Res.* **75**: 471–488.

Knudsen AG (1971) Mutation and cancer: Statistical study of retinoblastoma. *Proc. Natl. Acad. Sci. USA.* **68**: 620–623.

Knudsen AG (1985) Hereditary cancer, oncogenes and anti-oncogenes. *Cancer Res.* **45**(4): 1437–1443.

Knudsen AG (1991) Overview: Genes that predispose to cancer. *Mutat. Res.* **247**(2): 185–190.

Knudson AG, Meadows AT, Nichols WW, Hill R (1976) Chromosomal deletion and retinoblastoma. *N. Engl. J. Med.* **295**(20): 1120–1123.

Kobayashi H, Ohno S, Sasaki Y, Matsuura M (2013) Hereditary breast and ovarian cancer susceptibility genes (Review). *Oncol. Rep.* **30**(3): 1019–1029.

Konze-Thomas B, Hazard RM, Maher VM, McCormick JJ (1982) Extent of excision repair before DNA synthesis determines the mutagenic but not the lethal effect of UV radiation. *Mutat. Res.* **94**(2): 421–434.

Kraemer KH (1997) Sunlight and skin cancer: Another link revealed. *Proc. Natl. Acad. Sci. USA.* **94**(1): 11–14.

Kruuv J, Sinclair WK (1968) X-ray sensitivity of synchronized Chinese hamster cells irradiated during hypoxia. *Radiat. Res.* **36**(1): 45–54.

Lea DE, Catcheside DG (1942) The mechanism of the induction by radiation of chromosome aberrations in *Tradescantia. J. Genet.* **44**(2–3): 216–245.

Leenhouts HP (1999) Radon-induced lung cancer in smokers and non-smokers: Risk implications using a two-mutation carcinogenesis model. *Radiat. Environ. Biophy.* **38**(1): 57–71.

Leenhouts HP, Brugmans MJP (2000) An analysis of bone and head sinus cancers in radium dial painters using a two-mutation carcinogenesis model. *J. Radiol. Prot.* **20**(2): 169–188.

Leenhouts HP, Brugmans MJP, Chadwick KH (2000) Analysis of thyroid cancer data from the Ukraine after 'Chernobyl' using a two-mutation carcinogenesis model. *Radiat. Environ. Biophys.* **39**(2): 89–98.

Leenhouts HP, Chadwick KH (1974a) Radiation-induced DNA double strand breaks and chromosome aberrations. *Theor. Appl. Genet.* **44**(4): 167–172.

Leenhouts HP, Chadwick KH (1974b) The RBE-LET relationship. In: Booz J, Ebert HG, Eickel R, Waker A (eds.) *Fourth Symposium on Microdosimetry.* EUR 5122 (CEC, Luxembourg) pp 381–400.

Leenhouts HP, Chadwick KH (1976) Stopping power and the radiobiological effect of electrons, gamma rays and ions. In: Booz J, Ebert HG, Smith BGR (eds.) *Fifth Symposium on Microdosimetry.* EUR 5452 (CEC, Luxembourg) pp 289–308.

Leenhouts HP, Chadwick KH (1978a) An analysis of synergistic sensitization. *Br. J. Cancer.* **37**(Suppl. III): 198–201.

Leenhouts HP, Chadwick KH (1978b) The crucial role of DNA double strand breaks in cellular radiobiological effects. *Adv. Radiat. Biol.* **7**: 55–101.

Leenhouts HP, Chadwick KH (1983) Association between stochastic and non-stochastic effects and cellular damage. In: *Biological Effects of Low-Level Radiation. IAEA-SM-266/45* (IAEA, Vienna) pp 129–138.

Leenhouts HP, Chadwick KH (1984) A quantitative analysis of the cytotoxic action of chemical mutagens. *Mutat. Res.* **129**(3): 345–357.

Leenhouts HP, Chadwick KH (1985) Radiation energy deposition in water; calculation of DNA damage and its association with RBE. *Radiat. Prot. Dosim.* **13**(1–4): 267–270.

Leenhouts HP, Chadwick KH (1989) The molecular basis of stochastic and non-stochastic effects. *Health Phys.* **57**(Suppl. 1): 343–348.

Leenhouts HP, Chadwick KH (1994a) A two-mutation model of radiation carcinogenesis: Application to lung tumours in rodents and implications for risk evaluation. *J. Radiol. Prot.* **14**(2): 115–130.

Leenhouts HP, Chadwick KH (1994b) Analysis of radiation-induced carcinogenesis using a two-stage carcinogenesis model: Implications for dose-effect relationships. *Radiat. Prot. Dosim.* **52**: 465–469.

Leenhouts HP, Chadwick KH (1997) Use of a two-mutation carcinogenesis model for analysis of epidemiological data. In: *Health Effects of Low Dose Radiation: Challenges for the 21st Century* (British Nuclear Energy Society, London) pp 145–149.

Leenhouts HP, Chadwick KH (2011) Dose-effect relationships, epidemiological analysis and the derivation of low dose risk. *J. Radiol. Prot.* **31**(1): 95–105.

Leenhouts HP, Chadwick KH, Deen DF (1980) An analysis of the interaction between two nitrosourea compounds and X-radiation in rat brain tumour cells. *Int. J. Radiat. Biol. Relat. Stud. Phys. Chem. Med.* **37**(2): 169–181.

Leenhouts HP, Pruppers MJM, Chadwick KH (1990) Track structure, target structure and radiation effectiveness. *Radiat. Prot. Dosim.* **31**: 351–354.

Leenhouts HP, Sijsma MJ, Cebulska-Wasilewska A, Chadwick KH (1986) The combined effect of DBE and X-rays on the induction of somatic mutations in *Tradescantia. Int. J. Radiat. Biol. Relat. Stud. Phys. Chem. Med.* **49**(1): 109–119.

Leenhouts HP, Uit de Haag PAM, Chadwick KH (1996) Analysis of lung cancer after exposure to radon using a two-mutation carcinogenesis model. In: Goodhead DT, O'Neill P, Menzel HG (eds.) *Microdosimetry: An Integrated Approach* (Royal Society of Chemistry, Cambridge) pp 248–254.

Leuraud K, Richardson DB, Cardis E, et al. (2015) Ionising radiation and risk of death from leukaemia and lymphoma in radiation-monitored workers (INWORKS): An international cohort study. *Lancet Haematol.* **2**(7): e276–e281.

Lieber MR (2010) The mechanism of double-strand break repair by the non-homologous DNA end-joining pathway. *Ann. Rev. Biochem.* **79**: 181–211.

Linskens MHK, Huberman JA (1990) The two faces of higher eukaryotic DNA replication origins. *Cell* **62**(5): 955–965.

Little MP (2018) Evidence for dose and dose rate effects in human and animal radiation studies. *Ann. ICRP* **47**(3–4): 97–112.

Ljungman M, Hanawalt PC (1992) Localized torsional tension in the DNA of human cells. *Proc. Natl. Acad. Sci. USA.* **89**(13): 6055–6059.

Ljungman M, Nyberg S, Nygren J, Eriksson M, Ahnström G (1991) DNA-bound proteins contribute much more than soluble intracellular compounds to the intrinsic protection against radiation-induced DNA strand breaks in human cells. *Radiat. Res.* **127**(2): 171–176.

Lloyd DC, Edwards AA, Prosser JS (1986) Chromosome aberrations induced in human lymphocytes *in vitro* by acute X and gamma radiation. *Radiat. Prot. Dosim.* **15**(2): 83–88.

Lloyd DC, Edwards AA, Prosser JS, Corp MJ (1984) The dose response relationship obtained at constant irradiation times for the induction of chromosome aberrations in human lymphocytes by cobalt-60 gamma rays. *Radiat. Environ. Biophys.* **23**(3): 179–189.

Lloyd DC, Purrott RJ, Dolphin GW, Bolton D, Edwards AA,Corp MJ (1975) The relationship between chromosome aberrations and low LET radiation doses to human lymphocytes. *Int. J. Radiat. Biol.* **28**(1): 75–90.

Ludwików G, Xiao Y, Hoebe RA, et al. (2002) Induction of chromosome aberrations in unirradiated chromatin after partial irradiation of a cell nucleus. *Int. J. Radiat. Biol.* **78**(4): 239–247.

Maher VM, Dorney DJ, Mendrala AL, Konze-Thomas B, McCormick JJ (1979) DNA excision repair processes in human cells can eliminate cytotoxic and mutagenic consequences of ultra-violet irradiation. *Mutat. Res.* **62**(2): 311–323.

Major IR, Mole RH (1978) Myeloid leukaemia in X-ray irradiated CBA mice. *Nature* **272**(5652): 455–456.

Makalowski M (2003) Not junk after all. *Science* **300**(5623): 1246–1247.

Ma X, Liu Y, Liu Y, et al. (2018) Pan-cancer genome and transcriptome analyses of 1699 paediatric leukaemias and solid tumours. *Nature* **555**(7696): 371–376.

McKenna FW, Ahmad S (2011) Isoeffect calculations with the linear quadratic and its extensions: An examination of model-dependent estimates at doses relevant to hypofractionation. *J. Med. Phys.* **36**(2): 100–106.

McMillan TJ, Leatherman E, Ridley A, Shorrocks J, Tobi SE, Whiteside JR (2008) Cellular effects of long wavelength UV light (UVA) in mammalian cells. *J. Pharm. Pharmacol.* **60**(8): 969–976.

Metting NF, Braby LA, Roesch WC, Nelson JM (1985) Dose-rate evidence for two kinds of radiation damage in stationary-phase mammalian cells. *Radiat. Res.* **103**(2): 204–218.

Moan J, Porojnicu AC, Dahiback A (2008) Ultraviolet radiation and malignant melanoma. *Adv. Exp. Med. Biol.* **624**: 104–116.

Mole RH, Davids JAG (1982) Induction of myeloid leukaemia and other tumours in mice by irradiation with fission neutrons. In: Broerse JJ, Gerber GB (eds.) *Neutron Carcinogenesis* (CEC, Luxembourg) pp 31–42.

Mole RH, Major IR (1983) Myeloid leukaemia frequency after protracted exposure to ionizing radiation: Experimental confirmation of the flat-dose-response found in ankylosing spondylitis after a single treatment course with X-rays. *Leuk. Res.* **7**(2): 295–300.

Mole RH, Papworth DG, Corp MJ (1983) The dose-response for X-ray induction of myeloid leukaemia in male CBA/H mice. *Br. J. Cancer* **47**(2): 285–291.

Moolgavkar SH (1983) Model for human carcinogenesis: Action of environmental agents. *Environ. Health Persp.* **50**: 285–291.

Moolgavkar SH, Knudsen AG (1981) Mutation and cancer: A model for human carcinogenesis. *Proc. Natl. Acad. Sci. USA.* **68**: 1037–1052.

Mouret S, Baudouin C, Charveron M, Favier A, Cadet J, Douki T (2006) Cyclobutane pyrimidine dimers are predominant DNA lesions in whole human skin exposed to UVA radiation. *Proc. Natl. Acad. Sci. USA.* **103**(37): 13765–13770.

Mouret S, Philippe C, Gracia-Chantegrel A, et al. (2010) UVA-induced cyclobutane pyrimidine dimers in DNA: A direct photochemical mechanism? *Org. Biomol. Chem.* **8**(7): 1706–1711.

Muller HJ (1928)The production of mutations by X-rays. *Proc. Natl. Acad. Sci.* **14**(9): 714–726.

Murray D, Prager A, Milas L (1989) Radioprotection of cultured mammalian cells by the aminothiols WR-1065 and WR-2591: Correlation between protection against DNA double-strand breaks and cell killing after gamma-irradiation. *Radiat. Res.* **120**(1): 154–163.

Murray D, Prager A, Vanankeren SC et al. (1990) Comparable effect of the thiols dithiothreitol, cysteamine and WR-151326 on survival and on induction of DNA damage in cultured Chinese hamster ovary cells exposed to γ-radiation. *Int. J. Radiat. Biol.* **58**(1): 71–91.

Nagasawa H, Little JB (1981) Induction of chromosome aberrations and sister chromatid exchanges by X-rays in density-inhibited cultures of mouse 10T1/2 cells. *Radiat. Res.* **87**(3): 538–551.

NCRP (2018) Evaluation of the relative effectiveness of low-energy photons and electrons in inducing cancer in humans. NCRP Report No. 181.

Ochi T, Oshawa M (1983) Induction of 6-thioguanine resistant mutants and single-strand scissions of DNA by cadmium chloride in cultured Chinese hamster cells. *Mutat. Res.* **111**(1): 69–78.

Ohno S (1972) So much "junk" DNA in our genome. *Brookhaven Symp. Biol.* **23**: 366–370.

Ozasa K, Shimizu Y, Suyama A, et al. (2012) Studies of the mortality of atomic bomb survivors, Report 14, 1950–2003: An overview of cancer and non-cancer diseases. *Radiat. Res.* **177**(3): 229–243.

Park SD, Cleaver JE (1979a) Post replication repair: Question of its definition and possible alteration in *Xeroderma pigmentosum* cell strains. *Proc. Natl. Acad. Sci. USA.* **76**(8): 3927–3931.

Park SD, Cleaver JE (1979b) Recovery of DNA synthesis after ultraviolet irradiation of *xeroderma pigmenstosum* cells depends on excision repair and is blocked by caffeine. *Nucleic Acid. Res.* **6**(3): 1151–1159.

Peak JG, Peak MJ (1990a) Ultraviolet light induces double-strand breaks in DNA of cultured human P3 cells as measured by neutral filter elution. *Photochem. Photobiol.* **52**(2): 387–393.

Peak MJ, Peak JG (1990b) Hydroxyl radical quenching agents protect against DNA breakage caused by both 365-nm UVA and by gamma radiation. *Photochem. Photobiol.* **51**(6): 649–652.

Peak JG, Meyrick M, Peak MJ (1991a) Comparison of initial yields of DNA-to-protein cross-links and single-strand breaks induced in cultured human cells bt far- and near-ultraviolet light, blue light and X-rays. *Mutat. Res.* **246**(1): 187–191.

Peak MJ, Wang L, Hill CK, Peak JG (1991b) Comparison of repair of DNA double-strand breaks caused by neutron or gamma radiation in cultured human cells. *Int. J. Radiat. Biol.* **60**(6): 891–898.

Peng Y, Brown N, Finnon R, et al. (2009) Radiation leukemogenesis in mice: Loss of *PU.1* on chromosome 2 in CBA and C57BL/6 mice after irradiation with 1 GeV/nucleon ^{56}Fe ions, X-rays or γ-Rays. Part I. Experimental observations. *Radiat. Res.* **171**(4): 474–483.

Pierce DA, Shimizu Y, Preston DL, Vaeth M, Mabuchi K (1996) Studies of the mortality of atomic bomb survivors. Report 12, Part I. Cancer: 1950–1990. *Radiat. Res.* 146: 1–27.

Plaschka C, Hantsche M, Dienemann C, Burzinski C, Plitzko J, Cramer P (2016) Transcription initiation complex structures elucidate DNA opening. *Nature* **533**(7603): 353–358.

Platzman RL (1967) Energy spectrum of primary activations in the action of ionizing radiation. In: Silini G (ed.) *Radiation Research* (North Holland Publishing Company, Amsterdam) pp 20–42.

Poirier V, Papadopoulo D (1982) Chromosomal aberrations induced by ethylene oxide in a human amniotic cell line *in vitro. Mutat. Res.* **104**(4–5): 255–260.

Powers EL (1974) Is the water shell about the "target" involved in radiation effects in cells ? In: Booz J, Ebert HG, Eickel R, Waker A (eds.) *Fourth Symposium on Microdosimetry.* EUR 5122 (CEC, Luxembourg) pp 607–622.

Prise KM (1994) Use of radiation quality as a probe for DNA lesion complexity. *Int. J. Radiat. Biol.* **65**(1): 43–48.

Prise KM, Davies S, Michael BD (1987) The relationship between radiation-induced DNA double- strand breaks and cell kill in hamster V79 fibroblasts irradiated with 250 kVp X-rays, 2.3 MeV neutrons or ^{239}Pu α-particles. *Int. J. Radiat. Biol.* **52**(6): 893–902.

Prise KM, Davies S, Michael BD (1993) Evidence for induction of DNA double-strand breaks at paired radical sites. *Radiat. Res.* **134**(1): 102–106.

Prise KM, Folkard M, Davies S, Michael BD (1989) Measurement of DNA damage and cell killing in chinese hamster V79 cells irradiated with aluminum characteristic ultrasoft X rays. *Radiat. Res.* **117**(3): 489–499.

Prise KM, Folkard M, Davies S, Michael BD (1990) The irradiation of V79 mammalian cells by protons with energies below 2 MeV. Part II. Measurement of oxygen enhancement ratios and DNA damage. *Int. J. Radiat. Biol.* **58**(2): 261–277.

Prise KM, Gillies NE, Michael BD (1999) Further evidence for double strand breaks originating from a paired radical precursor from studies of oxygen fixation processes. *Radiat. Res.* **151**(6): 635–641.

Pruppers MJM, Leenhouts HP, Chadwick KH (1990) A track structure model for the spatial energy deposition of ionising radiation. *Radiat. Prot. Dosim.* **31**(1–4): 185–188.

Radford IR (1985) The level of induced DNA double-strand breakage correlates with cell killing after X-irradiation. *Int. J. Radiat. Biol. Relat. Stud. Phys. Chem. Med.* **48**(1): 45–54.

Radford IR (1986) Evidence for a general relationship between the induced level of DNA double-strand breakage and cell killing after X-irradiation of mammalian cells. *Int.J. Radiat. Biol.* **49**(4): 611–620.

Rao BS, Hopwood LE (1982) Modification of mutation frequency in plateau phase Chinese hamster ovary cells exposed to gamma radiation during recovery from potentially lethal damage. *Int. J. Radiat. Biol. Relat. Stud. Phys. Chem. Med.* **42**(5): 501–508.

Rapp A, Greulich KO (2004) After double-strand break induction by UV-A, homologous recombination and non-homologous end joining cooperate at the same DSB if both systems are available. *J. Cell Sci.* **117**(21): 4935–4945.

Resnick MA (1976) The repair of double-strand breaks in DNA: A model involving recombination. *J. Theor. Biol.* **59**(1): 97–106.

Resnick MA, Martin P (1976) The repair of double-strand breaks in the nuclear DNA of *Saccharomyces cerevisiae* and its genetic control. *Mol. Gen. Genet.* **143**(2): 119–129.

Revell SH (1966) Evidence for a dose-squared term in the dose-response curve for real chromatid discontinuities induced by X-rays and some theoretical consequences thereof. *Mutat. Res.* **3**(1): 34–53.

Revell SH (1974) The breakage-and-reunion theory and exchange theory for chromosomal aberrations induced by ionizing radiation: A short history. *Adv. Radiat. Biol.* **4**: 367–416.

Richardson DB, Cardis E, Daniels RD, et al. (2015) Risk of cancer from occupational exposure to ionising radiation: Retrospective cohort study of workers in France, the United Kingdom, and the United States (INWORKS). *Br. Med. J.* **351**: h5359.

Richold M, Holt PD (1974) The effect of differing energies on mutagenesis in cultured Chinese hamster cells. In: *Biological Effects of Neutron Irradiation* (IAEA, Vienna) pp 237–244.

Ritter MA, Cleaver JE, Tobias CA (1977) High-LET radiations induce a large proportion of non-rejoining DNA breaks. *Nature* **266**(5603): 653–655.

Rizzo JL, Dunn J, Rees A, Rünger TM (2011) No formation of DNA double-strand breaks and no activation of recombination repair with UVA. *J. Invest. Dermatol.* **131**(5): 1139–1148.

Roberts JJ, Pascoe JM, Plant JE, Sturrock JE, Crathorn AR (1971) Quantitative aspects of the repair of alkylated DNA in cultured mammalian cells, 1. The effect on Hela and Chinese hamster cell survival of alkylation of cellular macromolecules. *Chem.Biol. Interact.* **3**(1): 29–47.

Rossi HH (1990) The dose rate effectiveness factor. *Health Phys.* **58**(3): 359–361.

Roux JC (1974) Irradiation de Chlorelles par les rayonnements particuliers. Thèse, Université Scientifique et Médicale de Grenoble, CNRS A.O. 9896.

Rowland RE (1994) *Radium in Humans. A Review of U.S. Studies* (Argonne National Laboratories, Argonne, IL).

Rowley JD (1973a) A new consistent chromosomal abnormality in chronic myelogenous leukaemia identified by quinacrine fluorescence and Giemsa staining. *Nature* **243**: 290–293.

Rowley JD (1973b) Identification of a translocation with quinacrine fluorescence in a patient with acute leukaemia. *Ann. Genet.* **16**(2): 109–112.

Rühm W, Azizova TV, Bouffler SD, et al. (2016) Dose-rate effects in radiation biology and radiation protection. *Ann. ICRP* 45(no.1 suppl.): 262–279.

Rydberg B (1996) Clusters of DNA damage induced by ionizing radiation: Formation of short DNA fragments. II. Experimental detection. *Radiat. Res.* **145**(2): 200–209.

Rydberg B (2001) Radiation-induced DNA damage and chromatin structure. *Acta. Oncol.* **40**(6): 682–685.

Rydberg B, Löbrich M, Cooper PK (1994) DNA double-strand breaks induced by high energy neon and iron ions in human fibroblasts. I. Pulse-field electrophoresis method. *Radiat. Res.* **139**(2): 133–141.

Santiago A, Barczyk S, Jelen U, Engenhart-Cabillic R, Wittig A (2016) Challenges in radiobiological modeling: Can we decide between LQ and LQ-L models based on reviewed clinical NSCLC treatment outcome data? *Radiat. Oncol.* **11**: 67–80.

Sax K (1939) The Time factor in X-ray production of chromosome aberrations. *Proc. Natl. Acad. Sci. USA.* **25**(5): 225–233.

Sax K (1940) An Analysis of X-ray induced chromosomal aberrations in *Tradescantia. Genetics* **25**(1): 41–68.

Sax K (1941) Types and frequencies of chromosomal aberrations induced by X-rays. *Cold Spring Harbour Symp.* **9**: 93–103.

Scarpa A, Chang DK, Nones K, et al. (2017) Whole-genome landscape of pancreatic neuro-endocrine tumours. *Nature* **543**(7643): 65–71.

Schmid E, Rempl G, Bauchinger M (1974) Dose-response relation of chromosome aberrations in human lymphocytes after *in vitro* irradiation with 3 MeV electrons. *Radiat. Res.* **57**(2): 228–238.

Scudellari M (2017) To stay young, kill off the zombies. *Nature* **550**(7677): 448–450.

Searle AG (1974) Mutation induction in mice. *Adv. Radiat. Biol.* **4**: 131–207.

Serrano M (2017) Ageing: Tools to eliminate senescent cells. *Nature* **545**(7654): 294–296.

Shorrocks J, Paul ND, McMillan TJ (2008) The dose rate of UVA treatment influences the cellular response of HaCaT keratinocytes. *J. Invest. Dermatol.* **128**(3): 685–693.

Silver A, Moody J, Dunford R, et al. (1999) Molecular mapping of chromosome 2 deletions in murine radiation-induced AML localizes a putative tumour suppressor gene to a 1.0 cM region homologous to human chromosome segment 11p11-12. *Genes Chromosomes Cancer* **24**(2): 95–104.

Simpson P, Savage JRK (1996) Dose-response curves for simple and complex chromosome aberrations induced by X-rays and detected using fluorescence *in situ* hybridization. *Int. J. Radiat. Biol.* **69**(4): 429–436.

Sinclair WK (1966) The shape of radiation survival curves of mammalian cells cultured *in vitro*. In: *Biophysical Aspects of Radiation Quality*. Tech. Rep. Series 58 (IAEA, Vienna) pp 21–43.

Sinclair WK (1969) Protection by cysteamine against lethal X-ray damage during the cell cycle of Chinese hamster cells. *Radiat. Res.* **39**(1): 135–154.

Sinclair WK, Morton RA (1966) X-ray sensitivity during the cell generation cycle of cultured Chinese hamster cells. *Radiat. Res.* **29**(3): 450–474.

Spanjers WA, Engels FM, Werry PATJ, Chadwick KH, Leenhouts HP (1976) New evidence on chromosome rejoining after radiation. In: Booz J, Ebert HG, Smith BGR (eds.) *Fifth Symposium on Microdosimetry*. EUR 5452 (CEC, Luxembourg) pp 545–556.

Sparkes RS, Sparkes MC, Wilson MG, et al. (1980) Regional assignment of genes for human esterase D and retinoblastoma to chromosome band 13q14. *Science* **208**(4447): 1042–1044.

Stahl A, Levy N, Wadzynska T, Sussan JM, Jourdan-Fonta D, Saracco JB (1994) The genetics of retinoblastoma. *Ann. Genet.* **37**(4): 172–178.

Stratton MR, Campbell PJ, Futreal A (2009) The cancer genome. *Nature* **458**(7239): 719–724.

Sweigert SE, Rowley R, Warters RL, Dethlefsen LA (1988) Cell cycle effect on the induction of DNA double-strand breaks by X-rays. *Radiat. Res.* **116**(2): 228–244.

Thacker J (1979) The involvement of repair processes in radiation-induced mutation of cultured mammalian cells. In: Okada S, Imamura M, Terashima T, Yamaguchi H (eds.). *Radiation Research* (Toppan, Tokyo) pp 612–620.

Thacker J (1981) The chromosomes of a V79 Chinese hamster line and a mutant subline lacking HPRT activity. *Cytogenet. Cell Genet.* **29**(1): 16–25.

Thacker J, Cox R (1983) The relationship between specific chromosome aberrations and radiation-induced mutations in cultured mammalian cells. In: Ishihara T, Sasaki MS (eds.) *Radiation-Induced Chromosome Damage in Man* (A. R. Liss, New York) pp 235–275.

Thacker J, Cox R, Goodhead DT (1980) Do carbon ultrasoft X rays induce exchange aberrations in cultured mammalian cells? *Int. J. Radiat. Biol. Relat. Stud. Phys. Chem. Med.* **38**(4): 469–472.

Thacker J, Goodhead DT, Wilkinson RE (1983) The role of localized single-track events in the formation of chromosome aberrations in cultured mammalian cells. In: Booz J, Ebert HG (eds.) *Radiation Protection: Eighth Symposium on Microdosimetry*. EUR 8395 (CEC, Luxembourg) pp 587–595.

Thacker J, Stretch A (1983) Recovery from lethal and mutagenic damage during post-irradiation holding and low-dose-rate irradiations of cultured hamster cells. *Radiat. Res.* **96**(2): 380–392.

Thacker J, Stretch A, Stephens MA (1977) The induction of thioguanine-resistant mutants of Chinese hamster cells by γ-rays. *Mutat. Res.* **42**(2): 313–326.

Thacker J, Stretch A, Stephens MA (1979) Mutation and inactivation of cultured mammalian cells exposed to beams of accelerated heavy ions. II. Chinese hamster V79 cells. *Int. J. Radiat. Biol. Relat. Stud. Phys. Chem. Med.* **36**(2): 137–148.

Thacker J, Wilkinson RE, Goodhead DT (1986) The induction of chromosome aberrations by carbon ultrasoft X-rays in V79 hamster cells. *Int. J. Radiat. Biol. Relat. Stud. Phys. Chem. Med.* **49**(4): 645–656.

Thomas CA, Hamkalo BA, Misra DN, Lee CS (1970) Cyclization of eukaryotic deoxyribonucleic acid fragments. *J. Mol. Biol.* **51**(3): 621–632.

Thorne MC (2011) Comment on 'Radiation-induced cancer arises from a somatic mutation'. *J. Radiol. Prot.* **31**(3): 370–372.

Ting W, Schultz K, Cac NN, Peterson M, Walling HW (2007) Tanning bed exposure increases the risk of malignant melanoma. *Int. J. Dermatol.* **46**(12): 1253–1257.

Tisljar-Lentulis G, Henneberg P, Feinendegen LE, Commerford SL (1983) The oxygen enhancement ratio for single- and double-strand breaks by tritium incorporated in DNA of cultured T1 cells. Impact of the transmutation effect. *Radiat. Res.* **94**(1): 41–50.

Todd P (1967) Heavy-ion irradiation of cultured human cells. *Radiat. Res. Suppl.* **7**: 196–207.

Traynor JE, Still ET (1968). *Dose Rate Effect on LD50/30 in Mice Exposed to Cobalt-60 Gamma Irradiation* (USAF School of Aerospace Medicine, Brooks Air Force Base, TX) Rep. SAM-TR-68-97.

Tubiana M (2000) Radiation risks in perspective: Radiation-induced cancer among cancer risks. *Radiat. Environ. Biophys.* **39**(1): 3–16.

Ullrich RL, Jernigan MC, Satterfield LC, Bowles ND (1987) Radiation carcinogenesis: Time–dose relationships. *Radiat. Res.* **111**(1): 179–184.

Underbrink AG, Kellerer AM, Mills RE, Sparrow AH (1976) Comparison of X-ray and gamma-ray dose-response curves for pink somatic mutations in *Tradescantia* clone 02. *Radiat. Environ. Biophys.* **13**(4): 295–303.

Underbrink AG, Sparrow AH, Sautkulis D, Mills RE (1975) Oxygen enhancement ratios (OERs) for somatic mutations in *Tradescantia* stamen hairs. *Radiat. Bot.* **15**(2): 161–168.

UNSCEAR (2018) 2017 Report to the general assembly. *Annex B: Epidemiological Studies of Cancer Risk Due to Low Dose-Rate Radiation from Environmental Sources* (United Nations, New York).

Upton AC, Jenkins VK, Conklin JW (1964) Myeloid leukemia in the mouse. *Ann. NY Acad. Sci.* **114**: 189–202.

Van Leeuwen CM, Oei AL, Crezee J, et al. (2018) The alfa and beta of tumours: A review of parameters of the linear-quadratic model, derived from clinical radiotherapy studies. *Radiat. Oncol.* **13**(1): 91–102.

Venema LB, Leenhouts HP, Laheij GMH, Chadwick KH (1995) Use of a two-mutation model for the analysis of bone tumours induced by internal emitters: Implications for low dose risks. In: van Kaick G, Karaoglou A, Kellerer AM (eds.) *Health Effects of Internally Deposited Radionuclides: Emphasis on Radium and Thorium* (World Scientific, Singapore) pp 197–202.

Virsik RP, Harder D, Hansmann I (1977) The RBE of 30 kV X-rays for the induction of dicentric chromosomes in human lymphocytes. *Radiat. Environ. Biophys.* **14**(2): 109–121.

Virsik RP, Schäfer C, Harder D, Goodhead DT, Cox R, Thacker J (1980) Chromosome aberrations induced in human lymphocytes by ultrasoft Al_K and C_K X-rays. *Int. J. Radiat. Biol.* **38**: 545–557.

Virsik RP, Blohm R, Hermann KP, Harder D (1981) Fast short-ranged and slow long-ranged interaction processes involved in chromosoma aberration formation. In: Booz J, Ebert HG, Hartfiel HD eds. *Seventh Symposium on Microdosimetry.* EUR 7147 (CEC, Brussels) pp 943–955.

Vivek Kumar PR, Mohankumar MN, Zareena Hamza V, Jeevanram RK (2006) Dose-rate effect on the induction of HPRT mutants in human G_o lymphocytes exposed *in vitro* to gamma radiation. *Radiat. Res.* **165**(1): 43–50.

Wade MH, Lohman PHM (1980) DNA repair and survival in UV-irradiated chicken-embryo fibroblasts. *Mutat. Res.* **70**(1): 83–93.

Ward JF, Kuo I (1973) Deoxynucleotides-models for studying mechanisms of strand breakage in DNA. *Int. J. Radiat. Biol. Relat. Stud. Phys. Chem. Med.* **23**(6): 543–557.

Watson JD, Crick FHC (1953) Molecular structure of nucleic acids: A structure for deoxyribose nucleic acid. *Nature* **171**(4356): 737–738.

Watt DE, Al-Affan IAM, Chen CZ, Thomas GE (1985) Identification of biophysical mechanisms of damage by ionizing radiation. *Radiat. Prot. Dosim.* **13**(1–4): 285–294.

Watt DE, Kadiri LA (1990) Physical quantification of the biological effectiveness of ionizing radiations. *Int. J. Quantum Chem.* **38**(4): 501–520.

Weber KJ, Flentje M (1993) Lethality of heavy ion-induced DNA double strand breaks in mammalian cells. *Int. J. Radiat. Biol.* **64**(2): 169–178.

Wells RL, Bedford JS (1983) Dose-rate effects in mammalian cells. IV. Repairable and non-repairable damage in noncycling C3H 10T1/2 cells. *Radiat. Res.* **94**(1): 105–134.

Wheeler KT, Deen DF, Wilson CB, Williams ME, Sheppard S (1977) BCNU-modification of the *in vitro* radiation response in 9L brain tumor cells of rats. *Int. J. Radiat. Oncol. Biol. Phys.* **2**(1–2): 79–88.

Wischermann K, Popp S, Moshir S, et al. (2008) UVA radiation causes DNA strand breaks, chromosomal aberrations and tumorigenic transformation in HaCaT skin keratinocytes. *Oncogene* **27**(31): 4269–4280.

Yuan He, Chunh Yau, Jie Fang, et al. (2016) Near atomic resolution visualization of human transcription promoter opening. *Nature* **533**(7603): 359–365.

Yunis JJ, Ramsay N (1978) Retinoblastoma and subband deletion of chromosome 13. *Am. J. Dis. Child.* **732**(2): 161–163.

Zahao W, Steinfeld JB, Liang F, et al. (2017) BRCA1-BRD1 promotes RAD51-mediated homologous DNA pairing. *Nature* **550**(7676): 360–365.

Index